QUICK
SCIENTIFIC
TERMINOLOGY

More than 80 Self-Teaching Guides teach essential skills from management to math, computer programming to personal finance, writing to art appreciation.

Look for these and other popular STGs at your favorite bookstore!

Ashley, HUMAN ANATOMY

Dietrich/Wicander, MINERALS, ROCKS AND FOSSILS

Gordon, HOW TO SUCCEED IN ORGANIC CHEMISTRY

Houk/Post, CHEMISTRY: CONCEPTS AND PROBLEMS

Kuhn, BASIC PHYSICS

Kybett/Martin, DIGITAL ELECTRONICS

Kybett, ELECTRONICS, 2ND ED.

Ryan, BASIC ELECTRICITY, 2ND ED.

Moché, ASTRONOMY, 3RD ED.

QUICK SCIENTIFIC TERMINOLOGY

Kenneth Jon Rose

WILEY

A Self-Teaching Guide

John Wiley & Sons, Inc.

New York · Chichester · Brisbane · Toronto · Singapore

Publisher: Stephen Kippur
Editor: David Sobel
Managing Editor: Ruth Greif
Editing, Design, and Production: Publication Services, Inc.

This publication is designed to provide accurate and authoritative information in regard to the subject matter covered. It is sold with the understanding that the publisher is not engaged in rendering legal, accounting, or other professional service. If legal advice or other expert assistance is required, the services of a competent professional person should be sought. FROM A DECLARATION OF PRINCIPLES JOINTLY ADOPTED BY A COMMITTEE OF THE AMERICAN BAR ASSOCIATION AND A COMMITTEE OF PUBLISHERS.

Library of Congress Cataloging-in-Publication Data:

Rose, Kenneth Jon.
        Quick scientific terminology / Kenneth Jon Rose.
            p. cm.— (Wiley self-teaching guides)
        Includes index.
        ISBN 0-471-85763-7
            1. Science—Terminology—Programmed instruction.
    2. Science—Programmed instruction. I. Title. II. Series.
    Q179.R62 1988   88-725
    501'.4—dc19   CIP

Printed in the United States of America
88 89 10 9 8 7 6 5 4 3 2 1

To my brother, Seth
and my sister, Elizabeth
with love

To the Reader

Quick Scientific Terminology is designed to teach you how to build hundreds of scientific terms using a system of word parts. Whether you are studying for the Scholastic Aptitude Test (SAT) or the Graduate Record Examination, or you are planning to enter the scientific or health field, or whether you just want to increase your scientific vocabulary, this book is designed to help you. To use this book, you do not need a special background in the sciences, though a high school education (or the equivalent) will make the task much easier.

This is a self-teaching book; you work at your own speed. To be as effective as possible, the program is structured so that you actively participate in building new words from word parts. It requires that you pronounce the words that you build and that you understand their meanings.

To use this program with maximum effectiveness, you should proceed in the following way: Read the unit entitled "How to Build a Word" first so that you understand the basic structure of the program. Then continue on to read Units 1 and 2. These two units include word parts that are common to all scientific disciplines; you will be able to build hundreds of scientific terms just from the word parts included in these two units alone. Study them, and then go on to the unit or units that interest you.

Each unit is divided into numbered frames; each frame will ask you to write out a particular word or word part, although occasionally, each may ask a question that requires a brief answer. On the right side of the page you will find a reading column and on the left side a verification column. *Cover the verification column on the left side with a card or a slip of paper.* As you read through the frames, respond to the questions that are asked.

When you have written out your answer, move your card down to the next frame and verify your attempt. *Do not look at the answers in the left-hand column until you have genuinely tried to fill in the answer space.*

If you have misspelled a word, or have written an incorrect answer, be sure to understand why your answer was wrong, and write out the correct answer before continuing to the next frame. The answer in the left-hand column may also have a guide to the pronunciation of a word. Be sure to pronounce the new word you have formed aloud. If the answer column does not have a pronunciation for the new word, look up the word in your dictionary if you are not sure that you are saying it correctly.

Each frame consists of a statement or question and an answer space; if a word answer is required, the space will contain diagonals to separate the word parts (_____/_____/_____). If the question calls for a longer answer, you may use your own words to answer it.

At the end of each unit is a list of many of the words that you have built in that unit. There is also a self-test after each unit that is designed to test whether you have fully understood the words that you have built. If you complete the self-test and find that you have more than six incorrect answers, go back and review the unit again until the words and word parts are perfectly familiar to you.

At the back of the book are review sheets for each unit. You may go to them at any time and as often as you wish. They will help you keep the word parts fresh in your mind. In addition, the back of the book contains an index of word parts, which lists where you can find each word part in the body of the text. There is also a list of additional word parts, so that you may expand your scientific vocabulary even further.

In addition, two final self-tests are included. They will test the extent of your scientific vocabulary; from them you can ascertain how much you have learned.

So, enjoy!

K. J. R.

CONTENTS

Objectives of the Program

Upon completing *Quick Scientific Terminology*, you will have built more than 500 scientific terms using the word-building system, combining Greek and Latin prefixes, suffixes, word roots, and combining forms. You will have learned to:

1. identify and understand these word parts in your studies;
2. build other scientific terms from the word parts in this book; and
3. recognize new word parts in other scientific terms and use these word parts to build and understand further scientific terms, using this word-building design.

Pronunciation Key

The syllable bearing the primary stress or accent is indicated by capital letters, as in *ev-ery-man* (EV-rē-man).

a	add, map	m	move, seem	u	up, done
ā	ace, rate	n	nice, tin	er	urn, term
air	care, air	ng	ring, song	yoo	use, few
ä	palm, father	o	odd, hot	v	vain, eve
b	bat, rub	ō	open, so	w	win, away
ch	check, catch	ô	order, jaw	y	yet, yearn
d	dog, rod	oi	oil, boy	z	zest, muse
e	end, pet	ou	out, now	zh	vision, pleasure
ē	even, tree	oo	pool, food	ə	the schwa, an
f	fit, half	oo	took, full		unstressed
g	go, log	p	pit, stop		vowel represent-
h	hope, hate	r	run, poor		ing the sound
i	it, give	s	see, pass		spelled
ī	ice, write	sh	sure, rush		*a* in *above*
j	joy, ledge	t	talk, sit		*e* in *sicken*
k	cool, take	th	thin, both		*i* in *clarity*
l	look, rule	th	this, bathe		*o* in *melon*
					u in *focus*

Source: Modified "Pronunciation Key" in *Funk and Wagnalls Standard College Dictionary*. Copyright © 1977 by Harper and Row, Publishers, Inc. Reprinted by permission of the publisher.

In words like pneumonia and pneumatic, where pn begins a word, the "p" is silent; however, it may be pronounced when pn exists later in a word. Example: tach/y/pne/a.

In words like photic, phosphate, and philosophy, where ph begins a word, "ph" is pronounced like "f" as in fox or film.

How to Build a Word

Quick Scientific Terminology uses a system of word building that will teach you how to form thousands of scientific terms. Science is forever forming new words by borrowing the parts of other words. But, by following this system, you will be able to understand their meaning and their context.

word root

1. The whóle of a word is the sum of its parts. The foundation of a word is the word root. In the words im/plant, sup/plant, and trans/plant, plant is the _____ .

pack

2. What is the word root in these words: pre/pack, un/pack, pack/age? _____ .

cook

3. How about these: un/cook, pre/cook, cook/ing, cook/ed? _____ .

word roots

4. When two or more word roots are used to form a word, the term becomes a compound word. Compound words are formed from two or more _____ .

two words

5. Gum/shoe and life/guard are both compound words because both are formed from _____ .

madman

6. Build a compound word using the words mad and man: _____ .

moonquake

7. Build a term from the word roots moon and quake: _____ .

compound words

8. Both the words moonquake and madman are formed from two different words and are therefore _____ .

dynam/o

9. When a vowel is attached to a word root, the result is called a combining form. For the word dynam/o/meter, locate the combining form and write it here: _____/____ .

cry/o

10. What is the combining form for the term cry/o/stat? _____/____ .

retr
retr/o

11. In the compound words retr/o/fire, retr/o/grade, and retr/o/fit, the word root is _____ ; the combining form is _____/____ .

electr/o
positive

12. For the above terms, fire, grade, and fit are whole words. Many compound words are formed by using the combining form for the word root and attaching it to a whole word. In the term electr/o/positive what is the combining form? _____/____ . What is the whole word? _____ .

micr/o
balance

13. Try another. For micr/o/balance, what is the combining form? _____/____ . What is the whole word? _____ .

equ/i
molecul
-ar

14. In science, the majority of compound words are built up from a combining form, a word root, and an ending. In the term equ/i/molecul/ar, the combining form is _____/____ ; the word root is _____ ; the ending is _____ .

crystall/o/graph/y

15. Build a word from the combining form crystall/o, the word root, graph, and the ending -y.
_____ / / _____ / _____ .

atm/o/spher/ic

16. Try this variation. Build a term from the combining form atm/o, the word root spher, and the ending -ic: _____ / / _____ / _____ .

combining form
word root
ending

17. In the scientific term phot/o/graph/y, phot/o is the _____ ,
graph is the _____ ,
and -y is the _____ .

fixed
fixing

18. The ending that follows a word root is called the suffix. The endings -ed and -ing are suffixes. Attach them to this word root:
fix/_____ ;
fix/_____ .

one who views
closely

19. Using a suffix changes the meaning of the word. The suffix -or means one who. The word root inspect means to view closely. By adding the suffix -or (inspect/or) we get a word that means _____ .

suffix

20. The ending -less changes the meaning of the word effort in the compound effortless; -less is a _____ .

pre-
un-

21. A word part that precedes a word and thus changes its meaning is a prefix. In the words pre/cook and un/cook the prefixes are _____ and _____ .

un/like
dis/like

22. Using the prefixes dis- and un-, and the word root like, build two words that change the meaning of like: _____/_____ ; _____/_____ .

trans-
port
-er

23. In the term trans/port/er, the prefix is _____ ; the word root is _____ ; the suffix is _____ .

word root

24. To review: The foundation of the word is called the _____ .

combining form

25. When a vowel is added to a word root, the word part becomes a _____ .

compound word

26. When two or more word roots are combined to form a term, the term is called a _____ .

prefix

27. A word part that precedes a word root and changes the meaning of that word root is called a _____ .

suffix

28. A word part that follows a word root and changes the word root's meaning is called a _____ .

UNIT 1

Place, Direction, And Time

In Unit 1 you will form more than 100 scientific words by using the following word-root combining forms and prefixes, which describe place, direction, and time.

ab- *(away from)*
ad- *(toward)*
ante- *(before, forward)*
circum- *(around)*
de- *(down from, less than)*
dia- *(through)*
epi- *(over, upon)*
ex- *(from)*
extra- *(beyond)*
hyper- *(above)*
hypo- *(beneath)*
in- *(in, not)*
infra- *(under)*
inter- *(between)*
intra- *(within)*
mid- *(middle)*
para- *(around, near)*
per- *(through)*
peri- *(around)*
post- *(after, behind)*
pre- *(before)*
pro- *(before, in front)*

super- *(beyond)*
supra- *(above)*
sub- *(below, under)*
trans- *(across, over)*
ultra- *(above)*

anter/o *(before)*
caud/o *(tail)*
cephal/o *(head)*
dextr/o *(right)*
dors/o *(back)*
ect/o *(outer, outside)*
end/o *(within)*
ex/o *(outside)*
later/o *(side)*
lev/o *(left)*
medi/o *(middle)*
mes/o *(middle)*
poster/o *(behind, after)*
retr/o *(behind, backward)*
sinistr/o *(left)*
ventr/o *(belly)*

1. The word form dors comes from the Latin *dorsalis*, meaning back. If the word later/al means toward the side, the word dors/o/later/al must mean

toward the back
and sides

_____ .

dors/o/ventr/al
dor-sō-VEN-trəl

2. The combining form dors/o can also be combined with ventr/al, a word meaning toward the belly. Combine the two words to mean extending from the back to the belly: _____/____/_____ .

toward the belly
and the side

3. If ventr/al means toward the belly, and later/al means toward the side, then the word ventr/o/later/al must mean: _____ .

ventral
dorsal

4. On the human body, the navel is located on the _____ portion of the body (remember where the belly is located) and the spine is found in the _____ part.

medi/o

5. The navel is also located in the midregion or middle of the abdomen. A word that means middle is medi/al. The combining form for medial (medi/al) is _____/___ .

medi

6. Medi/al comes from the Latin word *medius*. Another word for middle is median. Both median and medial have the word root _____ .

dors/o/medi/al
dor-sō-MĒ-dē-əl

7. Combine the words dors/ and medi/al to mean towards the middle of the back: _____/____/_____ .

DIRECTIONAL WORD	COMBINING FORM	MEANING
dorsal	dors/al dors/o (back)	toward or on the back
ventral	ventr/al ventr/o (belly)	toward or on the belly side of the body

dorsomedial

8. The spine is in a _____ position in the body.

ventr/o/medi/al

9. The belly button, or navel, is in a _____/___/_____/_____ position.

mes/o/derm

10. The combining word form mes/o also means middle. Mes/o comes from the Greek *mesos*. Combine mes/o with the word derm, meaning skin: _____/___/_____ .

ect/o/parasite

11. There are three layers to an early embryo. The middle layer is the mesoderm, the outer layer is the ectoderm (ect/o/derm; look up ect/o), and the inner layer, or the layer within the embryo, is called the end/o/derm. Use one of these combining forms, ect/o, mes/o, or end/o, to define a parasite that lives on the *exterior* of its host: _____/___/_____ .

end/o

12. A plant that lives *within* another plant is a(n) (end/o, mes/o, ect/o) _____/___ /phyte.

end/o/enzyme

13. By the same token, an enzyme that functions within the confines of a cell must be a(n) _____/___/_____ .

midgut
midsection

14. Mid- also comes from the Greek *mesos*, and means middle. Midfield means the middle portion of a field. Midnight is the middle of the night. Midsummer is the middle of the summer. The word, then, meaning the middle of the alimentary canal or *gut* should be the _____ , and a *section* midway from the two ends is called the _____ .

DIRECTIONAL	COMBINING FORM OR PREFIX	MEANING
medial meso- mid-	medi/al medi/an mes/o mid-	middle (Latin) middle (Greek) middle (Greek)

15. For fun and a change of pace, look up these words in your dictionary and write a brief definition of each:

midbrain _____

_____

midrib _____

_____

mediator _____

mediate _____

_____

mesomorph _____

_____

mesosphere _____

_____

mesopause _____

_____

mesencephalon (mez-en-SEF-ə-län) _____

_____

16. The word anter/ior means toward the front or in front of; the word poster/ior means following or located behind. The combining forms of anterior and

anter/o
poster/o

posterior are _____/___ ;

_____/___ .

toward the front
middle of the
body

17. Something that is anter/o/medi/an is located

_____ .

poster/o/later/al

18. A word meaning something that is located toward the back and side of the body is ____/____/____ .

cephal/o
caud/o

19. The word cephal/ic means something that is toward the head, while the word caud/al means something that is located downward or toward the tail. Locate the combining forms for these two words and write them down here: ____/____ ; ____/____ .

TRUE. The stomach is in front of the backbone.

20. The stomach is anterior to the backbone. TRUE   FALSE (Are you covering the answers?)

TRUE. The feet are downward from the stomach.

21. The soles of the feet are caudal to the stomach. TRUE   FALSE

FALSE. The ears are located toward the sides of the body while the eyes are medial to the ears.

22. The eyes are lateral to the ears. TRUE   FALSE

Give the meaning or the word where indicated and then check your answers on the following page:

DIRECTIONAL WORD	MEANING
anterior	_____
_____	located behind, following
cephalic	_____
_____	downward, toward the tail

DIRECTIONAL WORD	COMBINING FORM	MEANING
anterior	anter/ior anter/o (before)	toward the front or in front of
posterior	poster/ior poster/o (behind, after)	following or located behind
cephalic	cephal/ic cephal/o	upward—toward the head
caudal caudad	caud/al caud/o /ad (tail)	downward—toward the tail

23. Label the correct positions on this drawing of a horse. See the left column for the answers.

1. dorsal
2. anterior (or cephalic)
3. posterior (or caudal)
4. cephalic (or anterior)
5. caudal (or posterior)
6. ventral

24. The prefixes supra-, ultra-, and hyper- all mean above or beyond, or excessive. For example, the word hyperactive is used to describe a person who is excessively active. By the same token, someone who is excessively *critical* is called

critical

hyper/_____ .

25. The Latin word for sound, *sonus*, is the basis for the adjective sonic. Write a word, using hyper-, that involves something going faster than the speed of

hyper/sonic

sound: _____/_____ .

hyper/ventil/ation
hī-pər-ven-tə-LĀ-
shən

26. A word for excessive breathing, or ventil/ation (from the Latin *ventilare*, to fan) above the normal is: _____ / _____ / _____ .

under or beneath

27. The opposite of hyper is hypo-. If hyper- means above, then hypo- must mean _____ .

beneath the skin
beneath the skin

28. If the word dermal means skin, then something that is hypo/derm/al must be something that lies _____ , and a hypodermic needle must be used to inject substances _____ .

hyperthyroid

29. The thyroid gland secretes chemicals called hormones (from the Greek *horman*, to stir up). A person who has a thyroid gland that secretes an excessive amount of hormones has a _____ condition.

hypo/thyroid

30. A person who possesses a thyroid gland that secretes an abnormally small amount of hormones has a _____ / _____ condition.

above, beyond,
or excessive

31. The prefix ultra-, like hyper-, means _____ .

ultra/liberal

32. A person who is extremely liberal is called _____ / _____ .

ultra/pure

33. So, a material that is extremely pure must be _____ / _____ .

ultra/son

34. And a sound having a frequency that is above the ear's ability to hear must be called _____ / _____ ic.

ultra/violet

35. By the same token, light that is beyond *violet* on the light spectrum (and is the cause of sunburns) should be called _____/_____ light.

infra

36. The opposite of above is below or under. The opposite of the prefix ultra- is the prefix infra-. So, for example, if ultraviolet light is on one end of the light spectrum, then _____ /red light should be on the other end, below red light.

above
beyond

37. Like ultra- and hyper-, the prefix supra- means _____ or _____ .

supra/spin/al
supraspinal

38. Something that is supracaudal (supra/caud/al) is above the tail. Something that is situated above the spine would be called _____/_____/___ . (Use the suffix -al, meaning of, or relating to.)

supra/natur/al

39. Something that transcends (or is beyond) the natur/al is called _____/_____/___ .

below or under

40. The opposite of supra- is sub-. If supra- means above or beyond, the prefix sub- must mean _____ .

sub/marine

41. A ship that operates below the ocean is a _____/_____ .

sub/standard

42. A material that is below the standard is _____/_____ .

sub/basement

43. A basement located below the true basement is the _____/_____ .

under the skin

44. The word cuticle comes from the Latin word *cutis*, meaning skin or hide. The cuticle in man is the outermost layer of skin. Therefore, something that is sub/cutaneous would be found _____ .

beneath

45. The cerebral (SER-ə-brəl) cortex (KOR-teks) is the outer layer of the brain responsible for all of our higher mental functions. The subcortex, then, would be that part of the brain immediately _____ the cortex. (Are you keeping the answers covered?)

sub

supra

ultra

hypo

hyper

sub

46. After choosing a matching prefix, look up these words in your dictionary and determine whether you made the correct choice. Then, write out a brief definition of the correct word in the spaces below.
(sub-) or (hypo-) _____ atomic

_____

(ultra-) or (supra-) _____ renal

_____

(supra-) or (ultra-) _____ centrifuge

_____

(hypo-) or (sub-) _____ physis

_____

(ultra-) or (hyper-) _____ bolic

_____

(infra-) or (sub-) _____ side

_____

super/charge
super/sonic

47. The prefix super- means over. Sometimes, though, it means above and on top of. If the words charge and sonic are each given the prefix super, they become _____/_____ ;
_____/_____ .

above the moon

48. The word lunar comes from the Latin for moon, *luna*. Thus, a body that is superlunar should appear _____ .

on or over
a continent

49. The prefix epi- has a number of meanings, including on, at, besides, over, outer, and after. For example, the epicenter (epi/center) is that part of the earth's surface directly *over* the center of an earthquake. Thus, an epi/continental sea must be a body of water that lies

_____ .

ex/o/skeleton

50. The combining form ex/o means outer, outside, or out of. For instance, the word ex/o/skeleton is the term for a skeleton that appears outside of the body. The hard shell of a lobster is the animal's

_____/____/_____ .

to cut out from

51. Unlike the combining form ex/o, the prefix ex- means from; more specifically, it means out from. For example, the verb exact (ex/act), comes from the Latin *exigere*, meaning to drive out from. The verb excise comes from the Latin *excidere*, which is the combination of ex- and *caedere* (to cut). Ex/cise, then, should have the meaning:

_____ .

from

52. To ex/hale is to remove air _____ the lungs.

ab-
away from

53. The prefix ab- also means from. However, its more specific meaning is away from. The words abduct, abhor, and ablate all have in common the prefix _____ , which means _____ .

to lead away from

54. The verb ab/duct comes from the Latin ab- + *ducere* (to lead). Therefore, the word ab/duct should mean _____ .

away from

55. An ab/duct/or muscle should be one that moves a
limb _____ the midline of the body.

ab/norm/al

56. The noun abnormal can be broken down into its
prefix, its combining form, and its suffix. Do so here:
_____/_____/_____ .

away from

57. Abnormal means something that is
_____ the normal.

ab-
ex-

58. The prefix de- means from, as well. In that, it
is similar to the two other prefixes, _____
and _____ .

away from
out from

59. As a reminder, the prefix ab- means
_____ ; the prefix ex- means
_____ . (Still covering the answers?)

down from

60. The more accurate definition of de- is down
from. The word de/mote means to be moved to a rank
_____ the previous one.

less than

61. De- also means from, or resulting in *less than*.
For example, the word de/hydr/ate has the prefix de-
and the combining form hydr. Hydr comes from the
Latin *hydr-*, meaning water. Thus, when water has
been removed from a substance, the substance has
_____ the water it had before.

de/hydr

62. Thus, when water is removed from a substance it
is ____/_____ /ated.

lacks water

63. When de/hydr/ation occurs in a cell, the cell
_____ .

toward

64. We now know that the prefix ab- means away from. The opposite of ab- is the prefix ad-. If ab- means away from, then ad- should mean _____ .

toward

65. The verb ad/duct means to draw _____ .

toward

66. You'll recall that the ab/ductor muscle moved a limb away from the midline of the body. Therefore, an ad/ductor must move a limb _____ the midline of the body.

caud/ad

67. Ad can also be a suffix, as in the word meaning toward the head, cephal/ad. A word meaning toward the tail, then, is _____/_____ .

ad/renal

68. If the word renal comes from the Latin word for kidney (*renes*), then a word for the gland that lies near (toward) the kidney should be _____/_____ .

69. A quick review before we go on: Fill in the meanings or prefixes where there are blanks. (Make sure to cover the answers.)

Prefix	Meaning
ab-	_____ away from
toward ad-	_____
ex-	_____ out from
ad-	_____ toward
de-	_____ down from
over, upon epi-	_____
under, below sub-	_____

around

70. Circum/navigate is a word that means to navigate, or sail, *around* (as the earth). A circum/lunar object, then, would be one that revolves _____ the moon.

around
around

71. Circum/volu/tion means a winding _____ a center. Circum/neutral means having a pH _____ neutral.

circum/oral

72. If a substance surrounded (were around) the mouth, or *oral* cavity, it would be a _____/_____ substance.

circum/polar

73. An object that remains above the earth's poles (use polar) is said to be a _____/_____ object.

around

74. Another prefix that means around is peri-. A peri/scope is a scope (Greek, to watch or spy) that allows the viewer to look _____ .

eye

75. A periscope is an optic/al instrument. That means that it involves the _____ . (Hint: what organ is needed to watch?)

around the eye

76. A material that is peri/optic is found _____ .

peri/spore

77. The word spore comes from the Greek *spora*, meaning seed. The material that covers or surrounds a spore is a _____/_____ .

around

78. In general, a word that has either peri- or circum- as its prefix usually means situated _____ an object.

through

79. The prefix per- is Latin and means through or throughout, and sometimes thoroughly. For example, the verb per/colate (from the Latin *colare*, to filter) means to cause a substance to pass _____ a filter.

through the skin

80. Recall that the word sub/cutaneous meant under the skin. As such, a per/cutaneous operation would be one performed _____ .

to complete
thoroughly

81. By the way, the verb per/form means, literally, _____ . (Hint: *founir*, to complete).

through
through

82. Per/for/ate (verb) means to make a hole _____ something. A per/for/ation is a hole _____ a substance.

through

83. The other prefix for through is dia-. The dia/meter (from the Greek *metron*, measure; pronounced dī-AM-ə-tər) is the length of a straight line passing _____ the center of a figure.

through

84. The word dia/lysis (French: *lyein*, to loosen) means the separation of substances by their unequal movement _____ a membrane.

through

85. A dia/pir (DĪ-ə-pir) is a fold of rock that has broken _____ the overlying layer of rock.

dia/treme
DĪ-ə-trēm

86. Build a word that means a volcanic vent through the earth's crust. (Hint: Use treme, from the Greek *trema*, hole.) _____/_____ :

pre-

87. "Do not pre/judge the pre/ced/ing paragraph." The common pre/fix in the last sentence is _____ .

before or in
front of

88. Logically, pre- should mean something that occurs _____ something else.

pre/atom
pre/mature

89. Build a word that means a time before the use of the atomic bomb: _____/_____ /ic. Build a word that means something that is mature before the proper time: _____/_____ .

ante/date

90. Another prefix meaning before is ante-. But ante-, unlike pre-, also means forward, and earlier than. A date of an event earlier than the actual date would be an _____/_____ .

ante/mort

91. The adjective mortal comes from the Latin word for death, *mortalis*. Form a word that means before death: _____/_____ /em. (Hint: Remove the -al from mortal.)

ante/nat
or pre/nat

92. The adjective natal comes from the Latin word for being born, *natalis*. Make a word that means before birth: _____/_____ /al.

behind
after

93. When an event occurs after an infant is born, it is said to be a post/natal event. The prefix post- has the opposite meaning to either pre- or ante-. The opposite of in front of is _____ ; the opposite of before is _____ .

post/mortem

94. An examination of a body after death is called a _____/_____ examination.

behind
after

95. Try these: Post/nasal means lying _____ the nose. Post/glaci/al means occuring _____ a period of glaciation (glā-shē-Ā-shən; from the Latin *glacies*, ice).

in front of
before
after birth

96. Now, these: Ante/colic means located _____ the colon. Ante/partum means _____ birth. Post/partum means _____ .

97. A few more:

before birth   Pre/natal means _____ ;

located forward   ante/rior means _____ ;

located behind   post/erior means _____ .

98. Like the prefix post-, the combining form retr/o means behind. It also means backward. The verb retr/o/gress (Latin: *gradi*, to go) means to move

backward   _____ .

99. A retr/o/reflect/or is a device that reflects light

backward   _____ .

100. However, a retr/o/lent/al (Latin: *lent*, lens) substance is a material that is situated

behind   _____ a lens.

101. The word serrate comes from the Latin *serratus*, saw. If a tree had serrated leaves, the edges of the leaves would resemble the toothed edge of a saw. If the tree had retr/o/serrat/ed leaves, the edges would be

backward   pointing _____ .

102. As a reminder, the two prefixes that mean in front of and before are: _____ and

ante-; pre-   _____ .

103. Another prefix that means in front of and before is pro-. Using pro-, build a word that means situated near the front of the head. (Hint: What is the word that means toward the head?)

pro/cephal   _____/_____ /ic.

104. When a protein (a complex molecule made of amino acids) is made by a cell, it is usually attached to a tail of amino acids. This tail makes the protein inactive. The tail must then be cut off for the protein to become active. Such an early protein is called a

pro/protein   _____/_____ .

105. An insulin molecule (a special protein) before the "tail" is cut off is called

pro/insulin   _____/_____ .

106. In this context, the prefix pro- means

before   _____ .

107. Actually, the proinsulin protein is part of an even larger molecule that must be cleaved off to yield the proinsulin molecule. This earlier version of proinsulin would be called
_____ /pro/insulin. (Hint: What prefix

pre   means "before"?)

108. The protein called glucagon (GLOO-kə-gän) is made in this way. List the three proteins in the process, in order of manufacture:

pre/pro/glucagon   _____/_____ /_____ ;
pro/glucagon   _____/_____ ;
glucagon   _____ .

Use the following information to work frames 109 through 119. This is another group of combining forms of direction and place.

COMBINING FORM	MEANING
dextr/o	to the right
lev/o	to the left
sinistr/o	to the left

right

109. In the preceding paragraph, the word glucagon is to the right of the prefix pro-. The combining form dextr/o means toward the _____ .

dextr

110. The hands of a clock viewed from in front rotate _____ /al/ly.

dextr/o/rotation

111. Build a word that means clockwise rotation: _____ / / _____ .

dextr/al

112. The spiral whorls of a seashell turn from left to right. That would make it a _____ / ____ shell.

to the left

113. The combining form that has the opposite definition of dextr/o is lev/o. If dextr/o means to the right, lev/o must mean _____ .

lev/o/rotation

114. Build a word that means a left-handed rotation: ____ / / _____ .

to or on the left

115. Lev/o is usually used only in molecular chemistry, while the other combining form, sinistr/o, is used elsewhere. Like lev/o, sinistr/o means _____ .

sinistr/al

116. A seashell that has spiral whorls that turn from right to left would be called a _____ / ____ shell.

dextr/o/manual

117. The word manual comes from the Latin *manualis*, hand. A right-handed person would be called a _____ / / _____ person.

sinistr/o/manual

118. A left-handed person would be called a
_____ / ___ / _____ person.

displacement
of the heart to
the right
displacement
of the stomach to
the left

119. Write out the meaning for each of the following
words:
dextr/o/cardi/a (cardi: heart) _____
_____
sinistr/o/gastr/ia (grastr: stomach) _____
_____

extending across
the Atlantic

120. The prefix trans- means across or through. What
is the meaning of the word trans/atlantic?
_____ .

trans/cutan

121. Build a word that means entering through
the skin. (What is the Latin word for skin?)
_____ / _____ /eous.

to carry across

122. The word trans/fer is a combination of trans-
and the Latin word meaning to carry, *ferre*. The literal
meaning of transfer is _____ .

trans/istor

123. A device for *trans*ferring an electrical
current across a res*istor* is called a
_____ / _____ .

trans

124. A protein capable of combining with ferric ions
(Latin: *ferrum*, iron) and transporting iron in the
body is called _____ /fer/rin.

trans

125. An enzyme that promotes the transfer of atoms
from one molecule to another is called:
_____ /fer/ase.

shining through

126. What is the literal meaning of the word trans/lucent? (Latin: *lucere*, to shine) _____ .

trans/mit

127. The verb trans/mit means to send across from one place to another. Build a word for a chemical released by nerves that travels from one cell to another: neur/o/_____/_____ /ter.

intra/galac

128. Intra- means inward or within. An event that occurs within the confines of a galaxy is _____/_____ /tic.

within a cell

129. Material that is intra/cell/ular exists _____ .

intra

130. Build a word that means occurring within the confines of the heart: _____ /cardiac.

between cells

131. The prefix inter-, unlike intra-, means between or among. Material that is inter/cell/ular exists: _____ .

inter

132. A neurotransmitter is an _____ /cellular chemical.

inter

133. The interval between the old and the new moon when the moon is invisible is called the _____ /lunar period.

inward
within

134. To review: Intra- means _____ or _____ .

between
among   **135.** Inter- means _____ or _____ .

Here are about 100 of the terms that you worked with in this unit. Pronounce each one carefully, then complete the Unit 1 Self-Test.

abductor	hyperdermal	prenatal
ablate	hypersonic	preproinsulin
adductor	hyperthyroid	procephalic
adrenal	hypodermal	proinsulin
antecolic	hypophysis	proprotein
antemortem	hypothyroid	retrogress
antenatal	intracardiac	retrolental
antepartum	intracellular	retroreflector
anteromedian	intragalactic	retroserrated
circumlunar	intercellular	sinistrogastria
circumnavigate	interlunar	sinistromanual
circumneutral	levorotation	subatomic
circumoral	mediate	subcortex
circumpolar	mediator	subcutaneous
circumvolution	mesencephalon	subside
dehydrate	mesoderm	superlunar
dextral	mesomorph	supracaudal
dextrocardia	mesopause	suprarenal
dextromanual	mesosphere	supraspinal
dextrorotation	midbrain	transcutaneous
dialysis	midgut	transferase
diapir	midrib	transferrin
diatreme	midsection	transistor
ectoderm	neurotransmitter	translucent
ectoparasite	percolate	transmit
endoderm	percutaneous	transmitter
endoenzyme	perforate	ultracentrifuge
endophyte	perioptic	ultrapure
epicenter	perispore	ultrasonic
epicontinental	posterolateral	ultraviolet
excise	postglacial	ventrolateral
exhale	postmortem	ventromedial
exoskeleton	postnatal	
hyperbolic	postpartum	

Unit 1 Self-Test

PART 1

From the list on the right, select the correct meaning for each of the following scientific terms:

____ 1. ventrolateral
____ 2. dorsomedial
____ 3. mesoderm
____ 4. midgut
____ 5. posterolateral
____ 6. caudal
____ 7. hyperventilation
____ 8. hyperthyroid
____ 9. supracaudal
____ 10. subcortex
____ 11. suprarenal
____ 12. excise
____ 13. abductor
____ 14. dehydrate
____ 15. circumvolution
____ 16. perioptic
____ 17. percutaneous
____ 18. antemortem
____ 19. postnasal
____ 20. procephalic
____ 21. dextromanual
____ 22. transferase
____ 23. intracellular

a. event occurring within cell
b. toward the side of the belly
c. enzyme that transfers
d. middle of the back
e. right-handed
f. near the front of the head
g. toward the back and the side
h. behind the nose
i. middle layer of the embryo
j. winding around a center
k. excessive thyroid gland
   secretion
l. found around the eye
m. middle of intestine
n. before death
o. toward the tail
p. moves a limb away from
   midline
q. above the tail
r. through the skin
s. beneath the cortex
t. to cut out
u. excessive breathing
v. above the kidney
w. removal of water
x. toward the head
y. below the tail
z. after death
aa. left-handed
bb. moves limb toward midline
cc. between cells

PART 2

Choose whether the following statements are TRUE or FALSE.

____ 1. Decay occurs in the antemortem body.
____ 2. The lungs are lateral to the arms.
____ 3. The outer layer of an embryo is the endoderm.
____ 4. A midsection of an apple includes the seeds.
____ 5. Neurotransmitters travel intracellularly.
____ 6. The earliest insulin is preproinsulin.

ANSWERS

Part 1

1. b	13. p
2. d	14. w
3. i	15. j
4. m	16. l
5. g	17. r
6. o	18. n
7. u	19. h
8. k	20. f
9. q	21. e
10. s	22. c
11. v	23. a
12. t	

Part 2

1. F   2. F   3. F   4. T   5. F   6. T

UNIT 2

Number, Size, and Color

In this unit you will put together more than 60 new scientific terms that indicate number, size, and color. Many (but not all) of the root combining forms and prefixes you will use are listed below.

bi-, di- *(two)*
demi-, hemi-, semi- *(half)*
multi- *(many, much)*
poly- *(many, several)*
chlor/o *(green)*
chrom/o *(color)*
chrys *(yellow, gold)*
cyan/o *(blue)*
erythr/o *(red)*

leuk/o *(white)*
macr/o *(large, long)*
meg/a *(great, large)*
melan/o *(black)*
micr/o *(small, minute)*
mon/o *(one, single)*
myri/o *(countless)*
null/i *(no, none)*
xanth/o *(yellow)*

Use the following chart of prefixes and combining forms to work frames 1 through 34.

NUMBER	MEANING	NUMBER	MEANING
uni-	one, single	hept/a	seven
bi-, di-	two, twice, double	oct/a	eight
tri-	three, thrice	non/a	nine
tetr/a	four, having four	dec/	ten
pent/a	five	undec/a	eleven
hex/a	six	dodec/a	twelve

one, single

1. Refer to the chart. The prefix uni- means
_____ .

single
one

2. A uni/cell/ular organism is a creature consisting of
a _____ cell. A uni/color/ous insect is one
that consists of _____ color throughout.

giving birth to one
offspring

3. Given that the adjective combining form
parous means giving birth to, what would
be the meaning of the word uni/parous?
_____ .

two, twice

4. The meaning of the prefix bi- is
_____ .

bi/alate
BĪ-ə-lāt

5. The adjective alate (Ā-lāt) means having wings. A
dragonfly is a _____/_____ creature.

two heads

6. A bi/cephal/ous creature would be one that has
_____ .

bi
BĪ-nōm-ən

7. In biology, organisms are given a two-part scientific
name. For example, the human species is called
Homo sapiens (wise man). Given that the word
nomen is Latin for name, *Canis latrans* is the
_____ /nomen of the coyote.

bi

8. A magnet, as well as the Earth, has two magnetic
poles. That would make each of these _____ /pol/ar
objects.

two, twice

9. The prefix di-, like bi-, also means
_____ .

two
methyl

10. A di/chloride molecule has _____ atoms of chlorine. A di/methyl molecule contains two _____ groups.

di/naphthyl
dī-NAF-thəl

11. Build a word that describes a molecule containing two naphthyl groups: _____/_____ .

tri

12. Build a word that describes a molecule containing three nitro groups: _____ /nitro.

tri/nomen

13. Build a word that means a three-part scientific name: ___/_____ .

tri/nomen

14. The Latin name for the sea cucumber, *Pawsonia saxi cola*, is a ___/_____ .

di/oxide

15. A tri/oxide is a compound containing three atoms of oxygen. Build a word that means a compound having two atoms of oxygen: ____/_____ .

four
four

16. Tetr/a is the combining form meaning _____ . A tetra/bromide contains _____ atoms of bromine.

tetr

17. Build a word that means a vacuum tube containing four electr*odes*: _____ /ode.

Pent/a

18. _____/___ is the combining form meaning five. (Look in the table.)

pent/a/chord

19. A violin is a four-stringed instrument, or a tetr/a/chord (Greek: *chordos*, stringed). A five-stringed instrument would be called a _____/__/_____ .

hex/a/chord

20. What should a guitar be called? (Hint: A guitar has six strings.) _____ / ___ / _____

21. A tri/atomic molecule is one that contains three atoms. Build words describing molecules that contain four, five, six, and seven atoms.

tetratomic   four atoms _____

pentatomic   five atoms _____

hexatomic   six atoms _____

heptatomic   seven atoms _____

22. The Pent/a/gon (Greek: *gonos*, angle) in Washington, D.C., is a building with five angles and therefore five sides. Build words for buildings having six, seven, and eight sides:

hexagon   six sides _____

heptagon   seven sides _____ _____

octagon   eight sides _____

hept/ode

23. The name for a vacuum tube containing seven electrodes is _____ / _____ .

seven molecules
of water

24. A hept/a/hydrate molecule is one containing how many molecules of what?

_____ _____ .

(octa)(deca)(di)ene

25. For the chemical octadecadiene, circle the combining forms indicating number.

eight
eight

26. How many limbs does an octopus have? _____ . Octandria are plants that all have flowers with _____ stamens.

nine

27. A non/a/gon (NÄN-ə-gän) has how many sides? _____ .

non/a/genarian

28. An oct/o/genarian (äk-tə-jə-NAIR-ē-ən) is an individual who is 80 or more but less than 90 years old. What would a person who is 91 be called? ____/____/_____ .

ten
oct/ə/pod

29. A lobster is a dec/a/pod (DEK-ə-päd). It has _____ limbs. (Latin: *poda*, foot.) An octopus is an _____/____/_____ .

dec/a
dek-ə

30. Build a word that means a compound with ten molecules of water: _____/____ /hydrate.

tri/dec

31. The combining forms that correspond to numbers after ten are easy to form. They are just combinations of combining forms. For example, the combining form for 11 is un/dec. What is the combining form for 13? _____/_____ .

hex/a/dec/a/
hydrate

32. Build a word that means a compound with 16 molecules of water: _____/_/____/_/_____ .

oct/a/dec/a/gon

33. Build a word that means an object having 18 sides: _____/_/____/_/_____ .

17

34. How many atoms does a hept/a/dec/atomic molecule have? _____ .

enlarging
(amplifying)

35. The combining form micr/o means small or minute. But it also means one millionth part of, or enlarging, as in the names of instruments, such as micr/o/scope. What is the meaning of micr/o in the word micr/o/phone? _____ .

a millionth of

36. What is the meaning of micr/o in the word micr/o/gram? _____ .

a millionth of

37. The rule is that with terms used in the metric system, like gram (a unit of weight) and meter (a unit of length), micr/o/ means a millionth of. What does micr/o mean in the word micr/o/liter (a unit of volume)? _____ .

a minute
meteorite

38. What does micr/o mean in the word micr/o/meteorite (meteor/ite [MĒT-ē-ə-rīt], a meteor that reaches the surface of the earth)? _____ .

micr/o/second

39. Build a word that means a millionth of a second: _____/ / _____ .

a million amperes

40. The combining form meg/a comes from the Greek word *megas*, meaning large, great, enhance. But it also means a million of. A meg/a/ampere (ampere [AM-pir], a unit of electrical current) implies _____ .

enhance

41. What does mega mean in meg/a/phone? _____ .

meg/a/lith

42. Using the combining form lith (stone), build a word that means huge stone: _____/ / _____ .

large or great
world

43. The combining form macr/o, like meg/a, means large, but it also means long, and often is used to indicate the opposite of micr/o. Micr/o/cosmos means little world. Macr/o/cosmos means _____ .

macr/o/cosm

44. One-celled organisms live in a micr/o/cosm. Planets and galaxies exist in the _____/___/_____ .

macr/o/climate

45. A micr/o/climate is the climate of a small geographic region. Build a word that means a climate of a large geographic region: _____/____/_____ .

large

46. A macr/o/dont has _____ teeth.

47. When a contract is null and void, it means that the contract has none of the binding power, or validity, that it had before. The combining form null/i, then, means _____ .

null/i/par

48. The word root par means give birth to (bear). A woman who has never borne offspring is called a _____/___/_____ /a (no births).

one

49. Mon/o means single or one. A mon/o/filament is a single filament. A mon/oxide is an oxide with _____ atom of oxygen.

one

50. A mon/o/polar magnet has _____ pole.

many

51. The opposite of one is many. Multi- means _____ .

many

52. Something that is multi/colored is made up of _____ colors.

mon/o

53. Nearly every cell has a nucleus. Those cells that have only one nucleus are _____/____ /nucle/at/ed.

multi

54. Those cells, like skeletal muscle cells, that have many nuclei are called _____ /nucle/at/ed cells.

none
one, single
many, much

55. To review again:
Null/i means _____ ;
mon/o means _____ ;
multi means _____ .

numerous

56. To build words meaning countless or numerous, myri/o is used. This combining form comes from the ancient Greek word *myrios*, meaning 10,000, a large number in those days. A myri/arch was a commander of 10,000 men in ancient Greece. From its name, the plant myri/o/phyllum (Greek: *phyllon*, leaf) has (few/numerous) _____ leaves.

numerous

57. Myri/o/nema is a plant that possesses _____ filaments.

myri/o/pod
MIR-ē-ə-päd

58. Build a term for an animal that has numerous appendages (use pod; Latin: feet).
_____ / _____ / _____ .

ten thousand

59. However, a myri/a/meter is a metric unit equal to _____ meters.

many

60. Poly- means many or several. In that, it is like the prefix multi-. Poly- is used to a greater extent. Poly/atomic molecules have _____ atoms.

many corners
(sides)

61. What does a poly/gon have?
_____ .

poly

62. Some animals have many mates at the same time. They are _____ /gamous animals.

many

63. A polymer is a chemical made up of _____ repeating structural units.

many or several

64. In general, then, poly- is used to indicate _____ .

Below is a chart that shows some of the prefixes of quantity. Use it while working frames 65 through 73.

PREFIX	MEANING	EXPLANATION
semi-	half	used most often with modern English words
hemi-	half	used with scientific terms
demi-	half	used least often

demi/circle
demi/sphere

65. Use the prefix demi- to form a word that means:
half circle _____ / _____ ;
half sphere _____ / _____ .

demi-

66. Demi- can connote something that is halfway between one thing and the other. A _____ /rhumb is the halfway point between rhumbs on the compass card.

semi/conscious

67. Semi- means half, but it can also mean incomplete or partial. A semi/coma is a partial coma. Someone who is half or partially conscious is _____ / _____ .

68. A figure that is partially erect is

semi/erect _____/_____ .

69. A material, such as a cell membrane, that is
partially permeable to substances in the environment

semi/permeable   is said to be _____/_____ .

70. Hemi- is strict in its meaning, which is

half   _____ .

71. Use hemi- to build a word that means:

hemi/section   a half section _____/_____ ;
hemi/sphere   a half sphere _____/_____ .

72. The cerebrum is a structure of the brain. A lateral
half of the cerebrum is a

hemi/cerebrum   _____/_____ .

73. A compound whose formula represents half that
of the chemical terpene is called

hemi/terpene   _____/_____ .

Use this information for building words involving quantity (frames 75
through 79):

PREFIX	MEANING	PREFIX	MEANING
deci-	tenth	deca-	ten
centi-	hundredth	hecto-	hundred
milli-	thousandth	kilo-	thousand
micro-	millionth	mega-	million
nano-	billionth	giga-	billion
pico-	trillionth	tera-	trillion
femto-	thousand trillionth	peta-	thousand trillion
atto-	million trillionth	exa-	million trillion

74. The prefixes in the chart above are used to indicate physical quantities that are either very large or very small. The prefixes on the left represent very small quantities, the ones on the right, very large quantities. A nano/second is a billionth of a second. This is a (very large/very small)

very small

_____ quantity.

milli/second
centi/meter
pico/watt
att/ohm
(look up
pronunciation in
your dictionary)

75. Using the prefixes for minute quantities, build a word that means:

a thousandth of a second _____/_____ ,

a hundredth of a meter _____/_____ ,

a trillionth of a watt _____/_____ ,

a million trillionth of an ohm _____/_____ .

tera/meter
kilo/second
mega/watt
hecto/volt
deca/meter
peta/second
(you look up)

76. Using the prefixes for very large quantities, build a word that means:

a trillion meters _____/_____ ,

a thousand seconds _____/_____ ,

a million watts _____/_____ ,

a hundred volts _____/_____ ,

ten meters _____/_____ ,

a thousand trillion seconds

_____/_____ .

a thousand
seconds
a thousand
trillion
meters

77. A kilo/meter is equal to a thousand meters. How many seconds equal a kilo/second?

_____ . How many meters

equal a peta/meter?

_____ .

deci/second
kilo/second
nano/meter
giga/watt

78. Build a word that means:

a tenth of a second _____/_____ ,

a thousand seconds _____/_____ ,

a billionth of a meter _____/_____ ,

a billion watts _____/_____ ,

hecto/liter	a hundred liters ———————/————— ,
pico/liter	a trillionth of a liter ———————/————— ,
micro/mole	a millionth of a mole ———————/————— ,
milli/mole	a thousandth of a mole ———————/————— .

79. Complete this chart and check your answers on the following page.

PREFIX	MEANING
pico- ———————	———————
atto- ———————	thousand
giga- ———————	thousandth
femto- ———————	billionth
micro-	million trillion
mega-	———————
tera- ———————	———————
peta- ———————	ten
deci-	hundredth ———————

Use this information to help you build terms involving color (frames 80 through 87):

COMBINING FORM	MEANING
leuk/o	white
melan/o	black
erythr/o	red
cyan/o	blue
chlor/o	green
xanth/o	yellow
chrys/o	yellow, gold

PREFIX	*MEANING*
pico-	trillionth
kilo-	thousand
atto-	million trillionth
milli-	thousandth
giga-	billion
nano-	billionth
femto-	thousand trillionth
exa-	million trillion
micro-	millionth
mega-	million
tera-	trillion
deca-	ten
peta-	thousand trillion
centi-	hundredth
deci-	tenth

green

80. Chlor/ine is a colored gas. From the combining form, what color is it? _____ .

cyan/osis

81. Chlor/osis is a disease that causes a greenish color of the skin. Form a word for a disease that would cause a bluish color of the skin: _____/_____ .

yellow

82. What color is the plant pigment called xanth/o/phyll? _____ .

melan/o/blast
leuk/o/blast
erythr/o/blast

83. Blast is a root word for an embryonic cell. Build a word that means an embyonic cell of the following colors:
black _____/__/_____ ,
white _____/__/_____ ,
red _____/__/_____ .

golden

84. Chrys/o/beryl is a colored mineral. From the combining form, what color is it? _____ .

85. What color is the beetle called chrys/o/melid?

golden   _____ .

86. Cyte means cell. Build a word that means:

erythr/o/cyte   red cell _____ / ___ / _____ ,

chlor/o/cyte   green cell _____ / ___ / _____ ,

leuk/o/cyte   white cell _____ / ___ / _____ ,

melan/o/cyte   black cell _____ / ___ / _____

green   **87.** Chlor/o means _____ ,

yellow   xanth/o means _____ ,

red   erythr/o means _____ ,

black   melan/o means _____ ,

yellow, gold   chrys/o means _____ ,

blue   cyan/o means _____ ,

white   leuk/o means _____ .

Below are 35 of the scientific terms you formed in Unit 2. Pronounce each one aloud before going on to Unit 3.

attohm	hectovolt	multicolored
bialate	hemicerebrum	myrionema
bicephalous	hemisection	nanosecond
centimeter	heptatomic	nullipara
chlorosis	hexachord	octadecagon
chrysoberyl	kilowatt	octagenarian
decahydrate	leukocyte	picowatt
demirhumb	macrocosm	polygon
dinaphthal	melanoblast	tetrode
erythroblast	micrometeorite	trinomen
exasecond	millisecond	unicolorous
gigawatt		

Unit 2 Self-Test

PART 1

From the list on the right, select the correct meaning for each of the following terms:

____ 1. microsecond
____ 2. heptagon
____ 3. melanocyte
____ 4. octadecagon
____ 5. gigawatt
____ 6. nullipara
____ 7. trinomen
____ 8. terameter
____ 9. erythroblast
____ 10. polygon
____ 11. macrocosm
____ 12. chlorosis
____ 13. heptadecagon
____ 14. leukocyte
____ 15. millisecond

a. no offspring
b. seventeen corners
c. three-part name
d. sixteen corners
e. white cell
f. thousand seconds
g. trillionth of a meter
h. seven corners
i. millionth of a second
j. eighteen sides
k. thousandth of a second
l. billion watts
m. early red cell
n. black cell
o. million seconds
p. large world
q. million watts
r. six corners
s. trillion meters
t. many sides
u. billionth of a watt
v. green skin

PART 2

Complete each of the following scientific terms on the right with the appropriate missing part:

1. blue skin   _____ osis
2. green cell   _____ cyte
3. thirteen sides   _____ gon
4. eight-part name   _____ nomen
5. sixteen sides   _____ gon
6. small world   _____ cosm

7. thousandth of a watt _____ watt
8. thousand watts _____ watt
9. million volts _____ volt
10. millionth of a liter _____ liter
11. twelve sides _____ gon
12. 10,000 meters _____ meter
13. five-part name _____ nomen
14. "great" bacterium _____ bacterium
15. billionth of a liter _____ liter

ANSWERS

Part 1
1. i
2. h
3. n
4. j
5. l
6. a
7. c
8. s
9. m
10. t
11. p
12. v
13. b
14. e
15. k

Part 2
1. cyanosis
2. chlorocyte
3. tridecagon
4. octanomen
5. hexadecagon
6. microcosm
7. milliwatt
8. kilowatt
9. megawatt
10. microliter
11. dodecagon
12. myriameter
13. pentanomen
14. megabacterium
15. nanoliter

UNIT 3

Basic Anatomy and Physiology

In this unit you will put together more than 100 new terms, using the following prefixes, word-root combining forms, and suffixes.

a- *(from, away)*
a-, an- *(less, not)*
contra- *(against)*
dis- *(apart, not)*
dys- *(difficult)*
em-, en- *(in, inside)*
eu- *(true)*
im- *(not)*

acr/o *(extremity)*
blast/o *(formative cell)*
brad/y *(slow)*
cardi/o *(heart)*
cerebr/o *(brain)*
cost/o *(ribs)*
crani/o *(skull)*
derm/o *(skin)*
gangli/o *(ganglia)*
gen/o *(generate)*
gon/o *(semen, seed)*
hem/o *(blood)*
hepat/o *(liver)*
hist/o *(tissue)*
lip/o *(fat)*
morph/o *(form)*

my/o *(muscle)*
ne/o *(new, recent)*
nephr/o *(kidney)*
neur/o *(nerve)*
oste/o *(bone)*
physi/o *(nature)*
pnea *(breath, air)*
pneum/o *(air, breath)*
ren/o *(kidney)*
splen/o *(spleen)*
tach/y *(rapid, speed)*
thorac/o *(chest)*
tonia *(tonic spasm)*
troph/o *(nutrition)*
trop/o *(turn, change)*
vas/o *(vessel)*

-ac *(pertaining to)*
-asis, -esis *(action, process)*
-ectomy *(cutting out)*
-gen *(something generated)*
-gram *(record, write)*
-graph *(instrument record)*
-ic *(of, relating to)*
-in *(belonging to)*

-ium *(small one, mass)*
-logist *(one who studies)*
-logy *(study of)*
-meter *(instrument for measuring)*
-oid *(like, in the form of)*
-on *(unit)*

-or *(one that does)*
-osis *(process, disease)*
-ous, -ose *(full of)*
-scope *(instrument for viewing)*
-stasis *(checking)*
-tomy *(cutting)*

cardi/o
KĂR-dē-ə

1. The word-root combining form for heart is
_____/_____ .

pertaining to
the heart
study of
the heart

2. Cardi/ac means
_____ .

Cardi/o/logy means
_____ .

within the mass
of the heart

3. End/o/cardi/um means
_____ .

around the mass
of the heart

4. Peri/cardi/um means
_____ .

tach/y/cardi
brad/y/cardi

5. Using the word root for heart, form a term
meaning: rapid heart _____/___/_____ __ /a;
slow heart _____/___/_____ /a.

cardi/o/blast

6. Using the word root for formative cell, build a
word meaning formative cell of the heart:
_____/___/_____ .

7. Build a word that means instrument that records
the heart:
_____/___/_____ .

cardi/o/graph
cardi/o/gram

Build a word that means tracing or record of the
heart: _____/___/_____ .

instrument for
viewing the
heart

8. What does the term cardi/o/scope mean?

_____ .

tach/o/meter

9. Build a term that means instrument for measuring
speed: _____ /o/_____ .

cardi/o/tach/o/
meter

10. Now build a word that means instrument
for measuring the speed of the heart:
_____ / / ____ / / _____ .

cardi/o/tomy

11. Build a word that means incision or cutting of
the heart:
_____ / / _____ .

my/o/cardium

12. Using the combining form variation
-cardium, build a word meaning heart muscle:
_____ / ____ / _____ .

formative muscle
cell

13. My/o/blast means

_____ .

my/o/cardi/o/graph

14. Build a word that means instrument that records
heart muscle: _____ / / ____ / / _____ .

my/o/graph
my/o/gram

15. Build a term that means: instrument that records
muscle activity_____ / / _____ ; record
of muscle activity_____ / / _____ .

the study of
muscles

16. What is my/o/logy?

_____ .

my/o/tonia

17. Build a word that means tonic spasm of muscle
_____/____/_____ .

muscle

18. My/osin, my/o/filaments, and my/o/globin are all proteins of _____ .

dys/pnea
DIS(P)-nē-ə
eu/pnea

19. A/pnea means not breathing. Build a word that means:
difficult breathing _____/_____ ;
normal or true breathing ____/_____ .

pneum/o/graph
pneum/o/gram

20. Pneum/o/logy (n[y]o͞o-MÄL-ə-jē) is the study of the respiratory organs: the lungs. Build a word that means instrument for recording respiratory movements: _____/___/_____ ; and a word that means record of respiratory movements: _____/___/_____ .

pneum/o/peri/cardi
/um

21. Build a word that means air in the peri/cardi/um: _____/___/_____/_____/ .

pneum/o/tach/o
/graph

22. Build a word for the device that records the speed of respiration: _____/___/_____/___ .

record of the speed
of respiratory
function

23. What does pneum/o/tach/o/gram mean? _____
_____ .

within the
circulation

24. An ex/o/crine gland is one that secretes its products outside of the circulatory system. Where does an end/o/crine gland secrete its products (hormones)?_____ .

formative cell
of nerves

25. Go to the list. Neur/o/blast means:

_____ .

26. Build an adjective that means a nerve that
secretes its products into the circulation:

neur/o/end/o/crine   _____/ /____/ /_____ .

27. End/o/crin/o/logy is the study of endocrine cells
which secrete their products into the circulation.
What is the term for the study of nerve cells that

neur/o/end/o/crin/o   secrete their products into the circulation?
/logy   _____/ /____/ /_____/ /_____ .

a hormone
secreted
by nerves

28. Define neur/o/hormone:

_____ .

29. The adjective for muscle is muscular. Build an
adjective that means relating to both nerves and

neur/o/muscular   muscles: _____/ /_____ .

study of the
activities
of nerves

30. Physi/o/logy is the study of the processes
and activities of life. Define neur/o/physi/o/logy:

_____ .

stopping blood
(arrest of bleeding)

31. Go back to the list. Hem/o/stasis means

_____ .

hem/o/peri/cardi
/um

32. Build a word that means blood in the pericardial
cavity: _____/ /_____/ /_____ .

blood

33. Hem/o/cyanin and hem/o/globin are pigments that
are both found in the _____ .

cerebr/o

34. Refer to the list again. Write the combining form that means brain: _____/_____ .

cerebr/al

35. The suffix -al makes a word an adjective. Make the word root for brain into an adjective: _____/_____ .

36. The term meaning half sphere is _____/_____ ;
Build a term that means either of

hemi/sphere
cerebr/al
hemi/sphere

the two halves of the brain (two words): _____/_____
_____/_____ .

cerebr/al

37. The cortex is the outer or superficial layer of an organ. What is the term for the outer layer of the brain (or cerebrum)? _____/_____ cortex.

cerebr/al

38. An artery that feeds the cerebrum is a _____/_____ artery.

crani

39. What is the word root that means skull? _____ .

crani/al

40. Using the suffix -al, build a word that means relating to the skull: _____/_____ .

crani/al
crani/al nerve

41. The bones of the skull are the crani/al bones. Build a term for the artery that passes through the skull: _____/_____ artery;
the nerve that passes through the skull: _____/_____ _____ .

crani/o/cerebr/al

42. Build a word that means relating to both the skull and the brain: _____/___/_____/_____ .

43. Build a word that means:
within the brain
_____/_____ /al;
within the skull
_____/_____/____ ;

intra/cerebr   within the nerve
intra/crani/al   _____/_____/___ ;
intra/neur/al   within the pericardium
intra/pericardi   _____/_____ /ac.

the study of   **44.** See the list. Hist/o/logy means
tissue   _____ .

                    45. Nerve, muscle, blood, bone, and the inner cell
                    lining of organs are tissues. Build a word that means
hist/o/blast   formative cell of tissue _____/___/_____ .

                    46. Hist/o/chemistry is the science that deals with the
tissues   chemicals of _____ .

                    47. Use these word parts to form a term that means
                    generating tissue:
                    hist/o
                    gen.
                    -ic
histo/o/gen/ic   _____/___/_____/_____

                    48. The Greek word *bryein* means
                    to swell. What is the literal meaning of
to swell inside   em/bryo?_____ .

                    49. The combining form for embryo is embry/o.
                    Embry/o/logy is the study of the development of an
                    individual from egg to birth. A person who studies
                    the development of an individual is an
embry/o/logist   _____/___/_____ .

embry/o/tomy

50. Build a word that means cutting or dissection of the embryo: _____/___/_____ .

of something generated

51. Gen/ic means _____ .

embry/o/genic

52. Carcin/o/genic means producing or generating cancer. Build a word that means generating an embryo: _____/___/_____ .

cost

53. The root word meaning rib is _____ .

between the ribs

54. Cost/al means pertaining to the ribs. Inter/cost/al means _____ .

inter/cost/al
in-ter-KÄS-təl

55. The muscles that move the ribs during breathing are called _____/_____/___ muscles.

intercostal

56. The nerves that supply these muscles are called _____ nerves.

oste/o

57. The combining form meaning bone is _____/___ .

oste/o/blast

58. Build a term for a formative cell that gives rise to bone: _____/___/_____ .

oste/o/crani/um

59. The bony skull is called the _____/___/_____/___ .

oste/o/gen/ic

60. Build a word that means generating bone: _____/___/_____/___ .

like bone   **61.** Oste/oid means _____ .

oste/o/tomy   **62.** Build a word that means cutting bone:
äs-tē-ÄD-ə-mə   _____/___/_____ .

                          63. The word root meaning form or shape is morph.
                          The form and structure (anatomy) of animals and
morph/o/logist   plants is studied by a _____/___/_____ .

                          64. Organ/o/gen/esis is the formation and
                          development of organs. The formation and
                          development of both organs and tissues is
morph/o/gen/esis   _____/___/_____/_____ .

                          65. The combining form for fat is lip/o. Build a word
                          meaning:
lip/oid   fatlike _____/_____ ;
lip/o/blast   formative fat cell _____/___/_____ ;
lip/o/gen/esis   formation of fat _____/___/_____/_____ ;
lip/osis   fat disease (obesity) _____/_____ .

                          66. Go back to the list. A nephr/o/logist studies the
kidney   _____ .

                          67. Form a word that means
                          formative kidney cell:
                          _____/___/_____ ;
                          like a kidney:
nephr/o/blast   _____/_____ ;
nephr/oid   kidney disease:
nephr/osis   _____/_____ .

                          68. Build a word that means kidney unit:
nephr/on   _____/_____ .

nephrons

69. Dissection of the kidney reveals that it is made up of millions of units called _____ .

ren/in
RĒN-ən

70. Another word root meaning kidney is ren, from the Latin *renes*. Form a word that means belonging to the kidney, for the enzyme the kidney secretes.
_____/_____ .

kidney

71. Ren/al is an adjective that means of or relating to the _____ .

ad/ren/al

72. Build a word that means toward or adjacent to the kidney: _____/_____/_____ .

toward the kidney

73. The ad/renal gland is found _____ .

hepat/o

74. What is the combining form that means liver? _____/____ .

hepat/o/gen/ic

75. Build a word that means producing or generating the liver: ____ _____/___/_____ /_____ .

cutting out

76. What does this combining form mean? -ectomy (Hint: ect/o + -tomy) _____ .

hepat/ectomy

77. Build a word that means cutting out the liver: _____/_____ .

hepat/o
meg/a
ly

78. Analyze hepat/o/meg/a/ly:
_____/____ combining form for liver;
_____/____ combining form for large;
_____ suffix.

hepat/o/meg/a/ly
enlargement of
the liver

79. Now put all the parts together:
_____/ / _____/ /_____ .
What does it mean?

_____ .

enlargement of the
spleen and liver

80. Analyze hepat/o/splen/o/meg/a/ly. What does
it mean? _____ .

enlargement of the
extremities (hands
and feet)
smallness of the
extremities

81. Define acr/o/meg/a/ly:

_____ ;

Define acr/o/micr/ia:

_____ .

vessel
vessel

82. Look back in the list. Vas/o means
_____ . A vas/o/spasm is a spasm of a
blood _____ .

83. Use the following verbs:
constrict—to make narrow
dilate—to spread wide
to build a term that means:
to make narrow a blood vessel

vas/o/constrict
vas/o/dilate

_____/ ____/ _____ ;
to spread wide a blood vessel
_____/ ____/ _____ .

84. The suffix -or makes a word a noun. Build a word
that means:
one that constricts blood vessels

vas/o/constrict/or
vas/o/dilat/or

_____/ ___/ _____/ __ ;
one that dilates blood vessels
_____/ / _____/ _____ .

vasoconstrictor

85. Vas/o/pressin is a hormone that causes
blood vessels to narrow. Vasopressin is a

_____ .

86. Thorac/ic (thə-RAS-ik) means relating to the chest.
The thoracic artery, nerve, and vein all supply the

chest   _____ .

87. Build a word that means:
instrument for viewing the chest cavity

thorac/o/scope   _____ / __ / _____ ;
thorac/o/tomy   cutting of the chest wall
thō-rə-KĀT-ə-mē   _____ / __ / _____ .

toward the back   **88.** Something that is thorac/o/dorsal is located
of the chest   _____ .

89. The Greek word for knot or gall is *ganglion*. A
ganglia/on is a mass of nerve tissue containing nerve
cells. What is the combining form for this term?

gangli/o   _____ / ____ .

90. Build a word that means:
gangli/o/blast   formative ganglion cell
gangli/on/ectomy   _____ / __ / _____ ;
gang-glēə-NEK-   cutting out (surgical removal) of a ganglion
tō-mē   _____ /on/_____ .

91. A vas/o/gangli/on is not a mass of nerve cells. It is
blood vessels   a knot of _____ .

92. Gon is the word root meaning semen or seed. A
gon/ad is a gland that produces (in the general sense)

seed   _____ .

93. Build a word that means cutting out the gonad:
gon/ad/ectomy   _____ / ____ / _____ .

nutrition
turning, or
stimulating

94. The word root troph means _____
vs. trop, which means

_____ .

stimulates

95. Adding the suffix -ic makes troph/ic, meaning
nourishing. But gonad/o/trop/ic hormone is a secretion
that _____ the gonad.

gonadotropic

96. LH, a hormone, stimulates the gonads. It is a
_____ hormone.

skin

97. As you learned from Unit 1, the combining form
derm/o means _____ .

derm/oid

98. Build a word that means resembling skin:
_____/_____ .

skin
skin

99. The combining form dermat/o also means
_____ ; a dermat/o/logist is one who studies
the _____ .

against conception

100. The word conception comes from the Latin
word, *concipere*, to become pregnant with.
Conception, then, is the act of becoming pregnant.
What is contra/ception?

_____ .

a newborn

101. The Latin *nasci* means to be born, and is related
to *natus*, born. What is a ne/o/nate?

_____ .

to cut apart

102. The Latin *secare* means to cut. What does
dis/sect mean? _____ .

dis/sect

103. To cut apart an organism is to do a
_____/_____ /ion.

full of poison

104. Poison/ous means _____ .

Below are 50 of the terms of anatomy and physiology that you formed in this unit. Read them one at a time and pronounce each aloud before you go on to the Unit Self-Test.

acromegaly	ganglionectomy	neonatal
acromicria	gonad	nephroblast
adrenal	gonadectomy	nephron
bradycardia	gonadotrophic	neuroblast
cardiotachometer	hepatectomy	neuroendocrine
cerebral	hepatogenic	neuromuscular
contraception	hepatomegaly	osteoblast
cranial	histoblast	osteogenic
craniocerebral	histogenic	osteotomy
dermatologist	intercostal	pneumopericardium
dermoid	intraneural	tachycardia
dissection	lipogenesis	thoracodorsal
dyspnea	liposis	thoracotomy
embryotomy	morphogenesis	vasodilator
endocardium	morphologist	vasoganglion
eupnea	myocardium	vasospasm
ganglioblast	myotonia	

Unit 3 Self-Test

PART 1

From the list on the right, select the correct meaning for each of the following terms. Write the letter in the space provided.

_____ 1. endocardium
_____ 2. bradycardia
_____ 3. myotonia
_____ 4. eupnea
_____ 5. neuroendocrine
_____ 6. intraneural
_____ 7. histogenic
_____ 8. osteoblast
_____ 9. lipogenesis
_____ 10. hepatectomy
_____ 11. acromegaly
_____ 12. vasodilator
_____ 13. gonadotropic
_____ 14. dissection
_____ 15. morphogenesis

a. nerve secretion into circulation
b. fast heartbeat
c. generating tissue
d. formative cell of bone
e. formation of organs and tissue
f. cutting out liver
g. one that widens vessels
h. cut apart
i. generating chest fat
j. around the heart
k. within the mass of the heart
l. tonic spasm of muscle
m. formation of fat
n. within the nerve
o. formation of bone
p. slow heartbeat
q. normal breathing
r. stimulating seed organ
s. enlargement of extremities

PART 2

Complete each of the terms on the right with the appropriate word parts:

1. around the heart   peri _____ um
2. slow heart   _____ ia
3. cutting of the heart   cardi _____
4. not breathing   _____ pnea
5. formative cell of nerves _____ blast
6. stopping blood   _____ stasis
7. relating to brain   _____ al
8. within the skull   inter _____ al
9. dissection of embryo   embryo _____

10. between the ribs   inter _____ al
11. generating bone   _____ gen _____
12. kidney unit   _____ on
13. towards the back of chest   _____ dors _____
14. formative "mass" or "knot" cell   _____ blast
15. resembling skin   _____ oid

ANSWERS

Part 1

1. k
2. p
3. l
4. q
5. a
6. n
7. c
8. d

9. m
10. f
11. s
12. g
13. r
14. h
15. e

Part 2

1. pericardium
2. bradycardia
3. cardiotomy
4. apnea
5. neuroblast
6. hemostasis
7. cerebral
8. intracranial

9. embryotomy
10. intercostal
11. osteogenic
12. nephron
13. thoracodorsal
14. ganglioblast
15. dermoid

UNIT 4

Botany

In this unit you will add new combining forms to the many you have already been introduced to in the preceding sections. Many of the new word parts you will use are in the following list:

andr/o *(man)*
anth/o *(flower)*
carp/o *(fruit)*
chlor/o *(green)*
chrom/o *(color)*
clad/o *(a shoot)*
coen/o *(shared in common)*
cole/o *(sheath)*
coni/o *(dust)*
herb *(grass)*
lign/i *(wood)*
phello *(cork)*

phot/o *(light)*
phyc/o *(seaweed)*
phyll *(leaf)*
phyt/o *(plant)*
plast *(formed)*
rhiz/o *(root)*
sapr/o *(rotten)*
scler/o *(hard)*
spor/o *(seed, spore)*
sym-, syn- *(together with)*
zyg/o *(yolk)*

plant life

1. The word, botany, comes from the Greek word *botanikos*, meaning of herbs. Today, botany deals with plants. Thus, botany is the study of _____ .

plant life

2. A botan/ist is one who studies _____ .

carp/o

3. The combining form for fruit is _____/_____ .

The following diagram shows the basic parts of a flower.

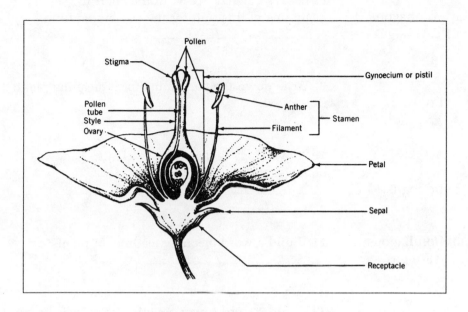

4. By looking at the combining form for fruit, you can tell that the structure of the flower called the carp/el (karpel) is involved in the production of

fruit   _____ .

5. The Greek word meaning producing is *gonos*. What is the literal meaning for the term carp/o/gonium?

_____ .

(Look up the word in your dictionary to get the exact

fruit producing   definition.)

6. Build a word meaning the study of fruit:

carp/o/logy   _____ / __ / _____ .

7. The adjective combining form -phagous (-fə-gəs) comes from the Greek word *phagos*, meaning one that eats. A cre/o/phagous organism is one that eats flesh. Build a word that means feeding on fruit:

carp/o/phagous   _____ / __ / _____ .

sapr/o/phagous
sa-PRAF-ə-gəs

8. Build a word that means feeding on rotten or decayed matter. (Look in the chart in the beginning of this unit for the combining form.)
_____ / / _____ .

phyt/o

9. Write down the combining form meaning plant:
_____ / ____ .

phyt/o/log/ist

10. Phyt/o/logy is the study of plants. And, if a botan/ist studies plants, so does a
_____ / / / _____ .

phyt/o/phagous
fī-TĂF-ə-gəs

11. Build a word meaning feeding on plants:
_____ / / _____ .

sapr
sapr/o/phyt/e
SAP-rə-fīt

12. Build a word for an organism that feeds on rotting plant matter: _____ /o/phyt/e. A mushroom feeds on rotting plant matter. A mushroom is a _____ .

phyt/o/plankton

13. The Greek word for wandering is *planktos*. Plankton (PLANK-tən) are passively drifting organisms of the sea. Zoo/plankton are passively drifting animals. Build a word for plant plankton:
_____ / / _____ .

phyt/o/benthon

14. *Benthos* is the Greek word for depths of the sea. Aquatic plants that live on the sea floor are called
_____ / / _____ . (Remove the -s in benthos and replace with -n.)

phyt/o/toxin

15. A toxin produced by a plant is a
_____ / / _____ .

16. The combining form phyc/o, comes from the Greek word *phycos*, and means seaweed. In botany, phyc/o is used for the plants called algae. Build a word that means the science of the study of algae:

phyc/o/logy _____/___/_____ .

17. As you may recall, the word root meaning red is erythr. Build a word for the red pigment occurring in

phyc/o/erythr red algae. _____/___/_____ /in.

18. The word root for blue is _____ . (See

cyan Unit 3 if you've forgotten.)

19. Build a term for the blue pigment occurring in

phyc/o/cyan blue-green algae. _____/___/_____ /in.

20. Bile is a fluid secreted by the liver that aids in digestion. The combining form for bile is bili. Bili/rubin (Latin: *ruber*, red) is a reddish pigment in bile. Bili/verdin (French: *verd*, green) is a green pigment in bile. Algae do not have bile, but have

bilelike pigments bilelike pigments. Give a definition of phyc/o/bili/n:
in algae _____ .

21. The Greek word for body is *soma*. Identify the three word parts in this term for bodies of pigment

phyc/o in blue-green algae: phyc/o/bili/some/s
bili some _____/_____ _____ _____ .

22. The discs of pigment attached to membranes in

phyc/o/bili/some/s blue-green algae are
fī-kō-BIL-ə- sōmz _____/___/_____ /_____ /_____ .

23. The Greek word for mushroom or fungus is *mykes*. The combining form for fungus is myc/o. Myc/o/logy is the science of the study of fungi (the plural of fungus). Build a word meaning feeding

myc/o/phagous   on fungi: _____/____/_____ .

mycophagous   **24.** Nematodes feed on fungi. They are
mī-KAF-ə-gəs   _____ creatures.

fungi or   **25.** A myc/o/phag/ist is one that eats
mushrooms   _____ .

                  26. Build a word for a toxin produced by a fungus:
myc/o/toxin   _____/___/_____ .

                  27. Identify the two word roots in the following
                  term for a class of fungi that are similar in
phyc   several characteristics to algae: phycomycetes
myc   _____ _____ .

                  28. The Greek word for cork is *phellos*. Cork is the
                  tough outer tissue of many trees. The combining form
phello   for cork is _____ .

                  29. You will recall that the Greek word for skin is
                  derma. Build a word for the layer of corklike
phello/derm   tissue formed in the stems and roots of some plants:
FELə-dərm   _____/_____ .

                  30. The Greek word for birth is *genesis*.
                  Give the literal meaning of the word phello/gen:
                  _____ . (Then look
gives birth to cork   the word up in your dictionary.)

phyll
fil

31. The word part for leaf is _____ . It comes from the Greek word for leaf, *phyllon*.

phyll/o/phagous

32. Phyll/oid means resembling a leaf. Build a word meaning feeding on leaves:

_____ /o/_____ .

phyllophagous
fil-ĂF-ə-gəs

33. Deer feed on leaves. Deer are _____ animals.

chlorophyll

34. Phyll can also occur in the latter part of a term. Chlor/o/phyll is the green pigment in leaves. Leaves are green because of the pigment _____ .

leaf

35. Mes/o/phyll is the middle layer of tissue in a _____ .

xanth/o/phyll

36. Recall that the Greek word for the color yellowish-brown is *xanthos*. The combining form is xanth/o. Build a word that means a yellow leaf (pigment): _____ /o/_____ .

spor/o

37. Write down the combining form for seed (spore): _____/____ .

spor/o/phyll

38. A spore is a reproductive cell that develops into a plant without the union of other cells. A spore-bearing leaf is called a _____ /o/_____ .

spor/o/phyte

39. What is the term for a spore-bearing plant? _____/____/_____ .

bears spores

40. The Greek word meaning to bear is *phorein*. The word part -phore is the root of phorein. A spor/o/phore is a tissue in fungi that _____ .

spore

41. Spore can also appear at the end of a term. Its meaning is the same. The term for the four spores formed by division of a spore mother cell is tetr/a/spores. The term contains the root word _____ .

zyg/o/spore

42. A spore developing from a zygote is a _____/_____/_____ . (Look up in word list.)

yolk

43. The root combining form zyg/o means _____ .

anth/o

44. Write down the combining form meaning flower: _____/_____ .

anth/o/phagous

45. The combining form anth/o comes from the Greek word *anthos*, flower. Anth/o/ecology is the study of flowers in their environment. Build a word meaning feeding on flowers: _____/_____/_____ .

anthophagous

46. Bees feed on flowers (or parts of them). Bees are _____ animals.

anth/o/phore

47. An organ of the plant that bears flowers is called an _____/_____/_____ .

anth/o/xanth

48. Build a term for the yellow-brown pigment found in flowers: _____/_____/_____ /in.

anth/o/cyan

49. Now build a word for the blue pigment in flowers: _____ / ___ / _____ /in.

anther

50. The root combining form anthera comes from the term for the pollen-bearing part of the stamen of the flower, the anther. The term anther/idium is made up of the root combining form anthera, which comes from _____ , and the suffix -idium, which means little one.

antheridium
anth-ə-RID-ē-əm

51. The term, then, for the pollen-bearing part of the plant is _____ .

clad/o

52. Write the combining form for shoot or sprout: _____ / ___ .

cladoptosis

53. Shoots often give rise to branches. Clad/o/ptosis (Greek: *ptosis*, dropping; klad-äp-TŌ-sis) is a term for the annual dropping of branches instead of leaves in some plants. The branches of the plants of the genus *Thuja* fall every year. These plants exibit _____ .

clad/o/phyll

54. Build a word that literally means shoot-leaf, a term for a branch that resembles a leaf: _____ / ___ / _____ .

herbs
herbs

55. The word herb (ərb) comes from the Latin word for grass, *herba*. In botany, the term is used for all seed plants that do not develop woody tissue. Trees and shrubs are not _____ . But grasses and weeds are _____ .

herbs

56. The term herb forms a part of several other scientific terms. The terms herb/aceous and herb/arium both refer to _____ .

57. The combining form -cide comes from the Latin word *cidere*, to kill. Build a word meaning a chemical that kills herbs (weed killer):

herb/i/cide

_____ /i/_____ .

58. The combining form -vorous comes from the Latin -*vorus*, eating. A carn/i/vorous animal is a meat-eating creature. What does an herb/i/vorous animal eat? _____ .

plants or herbs

59. A cow eats grass. That makes a cow an

herbivorous

_____ animal.

60. Write down the combining form meaning root:

rhiz/o

_____/___ .

61. The word rhiz/oid comes from the Greek *rhiza*, root, and the Latin *oides*, like. What does the term rhizoid mean? _____ .

rootlike

62. The term rhiz/anth/ous contains two word roots. Write their meanings here: _____ ;

root
flower
flower from
the root

_____ . Give a definition of the word, then look it up in your dictionary:

_____ .

63. The combining form meaning having the form is -morphous. Build a term that means having the form

rhiz/o/morphous

of a root: _____ /o/ _____ .

64. Build a word that means root-bearing:

rhiz/o/phore

_____/___/_____ .

65. The term gamete (gə-MĒT) comes from the Latin word *gamos*, marriage. In many animals and several plants, the sperm is the male gamete; the egg is the female gamete. Their marriage or union produces a fertilized egg. In botany, a gamet/o/phyte

plant

is a _____ that produces gametes.

66. Build a word meaning bearing gametes:

gamet/o/phore

_____ / ___ / _____ .

67. Recall that the combining form meg/a means large or great, and the combining form micr/o means small. In botany, meg/a is often associated with the female parts or reproductive cycle of many plants, and micr/o is associated with the male reproductive cycle. A meg/a/spore (large spore) gives rise to a female gamet/o/phyte. A micr/o/spore gives rise to a

male gametophyte

_____ .

a vessel that
contains spores

68. The Greek word for vessel is *angeion*. What is a spor/angium? _____ .

69. Build a word that means a sporangium that contains meg/a/spores.

meg/a/spor/angium

_____ / ___ / _____ / _____ .

70. Build a word that means a sporangium that contains micr/o/spores.

micr/o/spor/angium

_____ / ___ / _____ / _____ .

71. The Latin word for vessel is *cyta*. Give the literal definition of the term meg/a/spor/o/cyte:

large spore vessel

_____ .

micr/o/spor/o/cyte
mīk-rə-SPOR-ə-sīt

72. A megasporocyte is a cell that gives rise to four megaspores. Build a word for a cell that gives rise to four microspores: _____/ /_____/ /_____ .

large spore leaf

73. Give the literal definition of the term mega/spor/o/phyll: _____ .

micr/o/spor/o/phyll

74. A megasporophyll is a leaf that bears one or more megasporangia. Build a term meaning a leaf that bears microsporangia: _____/ /_____/ /_____ .

microsporophyll

75. The stamen of a flower bears microsporangia. A stamen is a _____ .

megasporophyll

76. A fern leaf bears megasporangia. A fern leaf is a _____ .

around

77. Many botanical terms can be formed using the prefix peri-. Peri- means _____ . (Look in Unit 2 if you've forgotten.)

peri

78. The peri/derm (Greek *derma*, skin) is the layer of cells around the inner layers of the tree. A term for the petals and sepals taken together is, literally, around flower, or _____ /anth.

peri/stome

79. Mosses have a fringe of "teeth" about the opening or mouth (Greek: *stoma*, mouth) of the sporangium. Build a word meaning around a mouth: _____/_____ . (Replace the -a with an -e.)

peri/carp

80. A fruit, like an apple, has three layers, the endo/carp, the meso/carp, and the peri/carp. Knowing what these prefixes mean, what would you call the skin of the apple? _____/_____ .

meso/carp

81. What would you call the flesh of the apple? _____/_____ .

endo/carp

82. The layer just surrounding the seeds of the apple is called the _____/_____ .

mes/o/phyte

83. What is the term for a plant that avoids both the extremes of moisture and drought (that takes the middle ground)? _____/__/_____ .

xer/o/phyte

84. The Greek word for dry is *xeros*. The combining form is xer/o. Build a term for a plant that lives in dry places: _____/__/_____ _____ .

xerophyte

85. A cactus is a _____ .

carp/y

86. Let's form a few botanical terms with the prefixes sym- and syn-. The two prefixes mean with or together with. The term for the condition where all of the petals on a flower are united (together with each other) is sym/petal/y. Build a term to mean a condition where all carpels are united: syn/_____/_____ .

syn/sepal/y

87. Build a term that means a condition where all of the sepals are united on a flower (use syn-): _____/_____/_____ .

syn/andr/y

88. The Greek word for man is *andros*. Stamens are the male part of the flower. So, with this in mind, build a word that means a condition where all of the stamens are united:
_____/_____/_____ .
(Use the word part, andr.)

thallus

89. Many plants begin as a thallus (THAL-əs; Greek *thallos*, sprout). A thallus is a plant body without true roots, stems, or leaves. A thall/o/phyte is a plant whose body is a _____ .

thallophytes

90. Algae and fungi have a body without true roots, stems, or leaves. They are _____ .

thall/o

91. The combining form in thallophyte that means thallus is _____/____ .

green

92. Write down the definition for the combining form chlor/o: _____ .

green leaf pigment

93. To refresh your memory, give the definition of the word chlor/o/phyll: _____ .

green
green

94. The combining form chlor/o means _____ . Special sacs within plant cells are called chlor/o/plasts (Greek: *plastos*, formed). From the name, chlor/o/plasts have a _____ color.

chlorophyll

95. Chloroplasts contain the pigment _____ .

96. Chloroplasts are often found in par/en/chyma
(Greek: *para*, around, *en*, in, *chein*, to pour; meaning
to pour in beside; pə-REN-kə-mə). Parenchyma is
an ancient Greek medical term reflecting the belief
that the tissues of the internal organs, like the liver,
were formed when blood poured into them and
coagulated. The par/en/chyma is a tissue of cells that
fit loosely together. Chloroplasts are often found in

par/en/chyma   leaf _____/___/_____ .

97. If the cells of parenchyma fit loosely together,
the cells of the coll/en/chyma (Greek: *kolla*, glue) fit

closely together   _____ .

98. Young stems contain a layer of cells that fit

collenchyma   closely together. This layer is _____ .

99. If you can remember it, write down the

chrom/o   combining form that means color: _____/____ .

100. Chrom/o/plasts contain yellow or orange
pigments. From their bright colors, the fall leaves

chrom/o/plasts   must contain _____/___/_____ .

101. The combining form meaning light is

phot/o   _____/____ .

102. Plants turn towards the light when they grow.
The Greek word for turning is *tropos*. The combining
form is trop/o. Build a word that means a turning to

phot/o/trop   the light: _____/___/_____ ism.

103. The Greek word for sun is *helio-*. The term

a turning toward   heli/o/trop/ism (noun) means
the sun   _____ .

putting together

104. The word syn/thesis (SIN-thə-sis) comes from two Greek word parts: *syn*, meaning together, and *tithenai*, to put. Literally, synthesis means _____ .

putting together
with light

105. Give a literal definition for the term phot/o/syn/thesis:
_____ .

phot/o/syn/thesis
fōt-ō-SIN-thə-sis

106. The process by which carbon dioxide and water are brought together chemically with the addition of light is called _____ / ___ / _____ .

plant
other

107. The combining form heter/o means other or different. Heter/o/sexual means of or relating to different sexes. A heter/o/phyte is a _____ that is dependent on _____ organisms for its food.

(two) different
spores

108. Heter/o/spor/y is the condition of producing _____ .

one and the same

109. In the condition of hom/o/spor/y, the same kind of spore is produced every time. Define the combining form hom/o: _____ .

heterothallic

110. Which term would best apply to the following definition: "referring to species in which male and female sex cells are produced by different plant bodies" ?
heterothallic
homothallic
_____ .

111. The term for the first leaf developed by the embryo of a seed plant is cotyledon (kät-l-ĒD-n; Greek: *kotyledon*, cup-shaped hollow). When there are two leaves, the term is di/cotyledon. What would be the term when there is one leaf?

mono/cotyledon _____/_____ .

112. The word part cotyl comes from the word

cotyledon meaning cup-shaped hollow, _____ .

113. You'll recall that hypo- means below or under. Build a word meaning that part of the plant embryo that is under the cotyledon (using the prefix hypo- and the combining form for cotyledon):

hypo/cotyl _____/_____ .

114. Build a word meaning literally "under the earth," a term for the type of germination where cotyledons remain below ground: _____/____ /al.

hypo/ge (Use the root ge, meaning earth.)

115. What is the combining form meaning hard?
_____/_____ . An instrument that determines

scler/o the hardness of materials is called a
scler _____ /o/meter.

116. Using the model term par/en/chyma, build a term

scler/en/chyma meaning plant tissue made of hard, thick-walled cells:
sklə-REN-kə-mə _____/__/_____ .

117. Coen/o/bium (Greek: *bios*, life) literally means life shared in common. The term is used to describe a colony of unicellular organisms surrounded by a common membrane. Certain algae exhibit

coen/o/bium this phenomenon. The colony, then, is a
sə-NŌ-bē-əm _____/__/_____ .

coen/o/cyte
SĒ-nə-sīt

118. An organism consisting of a mass of cellular fluid containing many nuclei within a single cell wall is a coen/o/cyte (Greek: *koinos*, shared in common, *kytos*, a vessel). An algae that contains this structure is a _____/__/_____ .

coen/o/gamete

119. Build a word for a gamete that is a single cell containing many nuclei:
_____/__/_____ .

cole/o

120. The combining form for sheath is
_____/____ .

cole/o

121. Cole/o/ptile (kō-lē-ÄP-tl; Greek: *ptilon*, down, feather) means the first leaf that sheaths the succeeding leaves. The combining form that lets you know that the leaf acts as a sheath is _____/___ .

cole/o/rhiz
kō-lē-ə-RĪ-zə

122. Build a word that literally means root sheath, but is the term for the sheath that surrounds the grass embryo and through which the roots burst:
_____/__/_____ /a.

coni/dia

123. Coni/o is the combining form that means dust. In botany it has come to mean spores, because the spores of fungi, when disturbed, blow away like dust. Coni/dia (kə-NID-e-ə) are asexual spores not borne within enclosing structures. Spores such as these found in fungi are called _____/_____ .

conidi/o/phore

124. Conidi/o is the combining form for conidia. Build a word for the structure that bears conidia:
_____ /o/ _____ .

conidi/o/spore
kə-NIDEə-spor

125. What is the term for a spore formed in conidia_____/__/_____ .

126. Lign/i is the combining form
that means wood. Like liqu/ify, which
to convert   means to convert into liquid, lign/ify means
into wood   _____ .

127. Lign/ite is a brown coal. From the name, you can
wood   tell that lignite came from _____ .

128. Using the prefix epi- (upon, over), build a word
meaning literally "upon the earth," a term for the
type of germination where cotyledons rise above the
epi/ge/al   ground: _____/_____/_____ .

129. Build a word meaning that part of the
plant embryo that is above the cotyledon (using the
prefix epi- and the combining form for cotyledon):
epi/cotyl   _____/_____ .

130. Build a word that means upon a plant, a term
that denotes a plant that grows upon another plant:
epi/phyte   _____/_____ .

In this unit you worked more than 90 new botanical terms. More than 80 of those words are listed here for you to practice your pronunciation. Do that first, then take the Unit Self-Test.

antheridium	chloroplast	dicotyledon
anthocyanin	chromoplast	endocarp
anthoecology	cladophyll	epigeal
anthophagous	cladoptosis	epiphyte
anthophore	coenobium	gametophore
anthoxanthin	coenocyte	gametophyte
botanist	coenogamete	heliotropism
carpogonium	coleoptile	heterophyte
carpology	coleorhiza	heterospory
carpophagous	collenchyma	heterothallic
chlorophyll	conidiophore	herbaceous

herbicide
herbivorous
hypocotyl
hypogeal
lignite
megasporangium
megasporocyte
megasporophyll
mesocarp
mesophyll
mesophyte
microsporangium
microsporophyll
monocotyledon
mycology
mycophagous
mycotoxin
parenchyma

perianth
pericarp
periderm
peristome
phelloderm
photosynthesis
phototropism
phycobilin
phycobilisomes
phycocyanin
phycoerythrin
phycology
phycomycetes
phyllophagous
phytobenthon
phytology
phytophagous
phytoplankton

phytotoxin
rhizoanthous
rhizomorphous
saprophagous
saprophyte
sclerenchyma
sporangium
sporophyll
sporophyte
synandry
syncarpy
synpetaly
synsepaly
tetraspores
thallophytes
xanthophyll
xerophyte
zygospore

Unit 4 Self-Test

PART 1

From the list on the right, select the correct meaning for each of the following often-used botanical terms.

____ 1. pericarp	a. feeding on decaying matter
____ 2. thallophyte	b. blue pigment in flowers
____ 3. parenchyma	c. root spore
____ 4. carpogonium	d. annual dropping of branches
____ 5. coleorhiza	e. loose layer of cells
____ 6. anthophagous	f. red pigment in algae
____ 7. homospory	g. root sheath
____ 8. saprophagous	h. gives birth to wood
____ 9. sporangium	i. skin of fruit
____ 10. phytophagous	j. fungi body without true stems
____ 11. cladoptosis	k. yellow leaf pigment
____ 12. phellogen	l. same kind of spore
____ 13. anthocyanin	m. feeding on flowers
____ 14. xanthophyll	n. fruit producing
____ 15. phycoerythrin	o. different kind of spore
	p. gives birth to cork
	q. vessel containing spores
	r. flesh of fruit
	s. feeding on plants

PART 2

Complete each of the botanical terms on the right with the appropriate prefix, suffix, or root combining form:

1. study of fruit   carp _____
2. plant plankton   _____ plankton
3. bodies of pigment in blue-green algae   _____ bili _____
4. one that eats fungi   Myco _____ ist
5. bears spores   sporo _____
6. branch resembling leaf   _____ phyll
7. rootlike   rhiz _____

8. large spore leaf   _____ sporo _____

9. plant living in desert   _____ phyte

10. united stamens   syn _____ y

11. plant dependent on different
      organisms for food   _____ phyte

12. under the earth   _____ ge _____

13. plant tissue of hard cells   _____ en _____

14. single-celled gamete with many
      nuclei   _____ gamete

15. to convert to wood   _____ ify

ANSWERS

Part 1

1. i	9. q
2. j	10. s
3. e	11. d
4. n	12. p
5. g	13. b
6. m	14. k
7. l	15. f
8. a	

Part 2

1. carpology	9. xerophyte
2. phytoplankton	10. synandry
3. phycobilisomes	11. heterophyte
4. mycophagist	12. hypogeal
5. sporophore	13. sclerenchyma
6. cladophyll	14. coenogamete
7. rhizoid	15. lignify
8. megasporophyll	

UNIT 5

Geology

In this unit you will form more than 70 geological terms. Some of the new combining forms and suffixes you will use are listed below.

all/o *(other)*
anis/o *(unequal)*
aqu/a, i *(water)*
aut/o *(same)*
bio *(life)*
calc/o *(lime)*
carb/o *(carbon)*
cry/o *(cold)*
crystall/o *(crystal)*
ferr/o *(iron)*
fluvi/o *(river)*
ge/o *(earth)*
glaci/o *(glacier)*
hyal/o *(glass)*
hydr/o *(water)*
lith/o *(stone)*
metall/o *(metal)*

morph/o *(form)*
pale/o *(old)*
ped/o *(soil)*
pel/o *(mud)*
petr/o *(rock)*
pseud/o *(false)*
pyr/o *(fire)*
seism/o *(earthquake)*
sider/o *(iron)*
silic/o *(silica)*
strat/i *(layer)*
uran/i *(uranium)*
xen/o *(strange)*

-fer *(one that bears)*
-iferous *(bearing)*
-ite *(mineral, rock)*

earth

1. The combining form, ge/o means _____ .

earth

2. Ge/o/logy is the science that deals with the _____ .

ge/o

3. Gold and oil are chemicals of the earth. The study dealing with the distribution of these chemicals is called ____/____ /chemistry.

ge/o/dynamic

4. The study of the dynamic forces within the earth is called _____/_____/_____ /s.

earth

5. A ge/ode (Greek: earthlike) is a nodule of stone with an inner cavity lined with crystals. From its name, a geode has something to do with the _____-__ .

ge/o/logist

6. Build a word that means a specialist in ge/o/logy: _____/_____/_____ .

ge/o/magnetic

7. Build a word that means pertaining to the magnetic field of the earth: _____/_____/_____
                (earth-magnetic)

earth

8. As a rule, then, any word that begins with the combining form ge/o pertains to the _____ .

Use these combining forms of the elements for frames 9 to 51:

COMBINING FORM	MEANING
aqu/a	water
calc/o	lime
carb/o	carbon
cupr/o	copper
ferr/o	iron
sider/o	iron
hyal/o	glass
metall/o	metal
silic/o	silica
uran/i	uranium

calc

9. Limestone is made up of calc/ium carbonate. Calc/ific/ation (kal-sə-fə-KĀ-shən) is the process of replacing the hard parts of a plant or animal with calcium carbonate. The word root for calc/ific/ation is _____ .

calcium carbonate or lime

10. Calc/aren/ite (kal-KAIR-ə-nīt), calc/ilut/ite, and calc/irud/ite (kal-sə-ROO-dīt), all contain the word root calc. Therefore, they must contain _____ .

rock or mineral

11. Calc/aren/ite and the other words above contain the suffix -ite. That makes each of them a _____ .

calc/ite
KAL-sīt

12. Build a word using the word root calc and the suffix -ite: _____/_____ .

a rock containing calcium carbonate

13. What does the word calc/ite mean literally? _____ .

like glass

14. The word hyal/ine (HĪ-ə-lin) comes from the Greek word *hyalinos*, meaning like glass. A hyaline object is one that is as clear or as smooth as glass. Hyal/ite is a variety of opal that can be described as being _____ .

hyal/o/basalt

15. Use the combining form hyal/o to build a word for a basalt (a dark grey to black fine- or dense-grained rock formed from lava) that has the texture of glass: _____/_/_____ .

partly glassy, partly crystalline

16. What would be the feel or texture of a rock that is described by the term hyal/o/crystall/ine? _____ .

crystal

17. The word crystalline contains the word crystal. A crystal is a solid body of a chemical element that has a regularly repeating atomic arrangment. A diamond is a _____ of carbon.

study of crystals

18. If ge/o/logy is the study of the earth, crystall/o/graphy is the _____ .

crystall/ite

19. Build a word that means a crystal mineral: _____/_____ .

the process of
becoming
crystallized

20. Hyalin/ization is a term that means the process of becoming hyalinized. What does crystall/ization mean? _____ .

aqu/a or aqu/i

21. Fish are aqu/a/tic creatures that can survive well in an aqu/a/rium. Write the combining form that means water: _____/_____ .

bears water

22. The suffix -fer means one that bears. Write the definition of an aqu/i/fer: A body of rock that _____ .

aqu
AK-wə-tärd

23. The verb retard means to slow up. A rock bed that retards or slows the movement of water is an _____ /i/tard. Write out the word: _____ .

cupr

24. Write out the word root for copper: _____ .

cupr/ite
KOO-prīt

25. Build a word for a mineral (rock) that is an ore of copper: _____/_____ .

bears copper

26. A mineral that is cupr/i/fer/ous
is one that _____ .
(Hint: What does fer mean?)

copper

27. The mineral cupr/o/nickel contains both nickel
and _____ .

mineral
min-ə-RÄL-ə-jē

28. The word mineral comes from the Latin word for
mine, *mineralis*. Build a word that means the study
of minerals: _____ /ogy.

iron

29. The combining form ferr/i means _____ .

ferr/i/fer/ous
fə-RIF-ər-əs

30. Build a word that means iron-bearing:
_____/____/_____/_____ .

ferr/i/magnet

31. An iron magnet is a
_____/____/_____ .

ferr/ite

32. A rock that contains mostly iron is called
_____/_____ . (Hint: What is the suffix
for rock?)

sider/ite
SID-ə-rīt

33. The word root ferr is used when describing
ferric iron. For ferrous iron, the word root sider is
used. Build a word for a ferrous iron-containing
mineral: _____/_____ .

sider/o/sphere

34. The word that means earth shell (the solid
earth) is ge/o/sphere. Build a word that means
the central iron shell or core in the earth:
_____ /o/_____ .

ferr

35. To review quickly: The word root _____ is used to mean ferric iron.

sider/o

36. The combining form _____/____ is used to mean ferrous iron.

replacement by silica

37. You will recall what calcification meant. What is the definition of silic/ific/ation? _____ .

silic

38. Write out the word root that indicates silica: _____ .

silica

39. Silic/eous shale (a hard, fine-grained rock) contains _____ .

rock

40. The combining form petr/o means _____ .

rocks

41. Petr/o/logy is the science that deals with the origin and structure of _____ .

petr/o/leum

42. *Oleum* is the Latin word for oil. What word has the literal meaning rock oil? _____ /o/_____ .

petr/o/glyph
PET-rə-glif

43. Hier/o/glyphs (Greek: *hieros*, sacred; *glyphikos*, carving) are found on the walls of Egyptian tombs. A carving on rock is called a _____/___/_____ .

stone or rock

44. Petr/i/fied wood is wood that has been replaced by _____ .

silica

45. Silic/i/fied wood is wood that has been replaced
by _____ .

uran/i

46. The combining form for uranium is
_____ / _____ .

uran/ium

47. Uran/in/ite (yoo-RĀ-nə-nīt) is a highly radioactive
mineral and the chief ore of _____ / _____ .

carbon

48. Carb/o/naceous (suffix -aceous, consisting of)
material is rich in the element _____ .

carb/o/naceous
kär-bə-NĀ-shəs

49. The word carbon, from which the combining
form carb/o derives, comes from the Latin *carbo*,
which means charcoal or ember. Build a word
meaning consisting of carbon:
_____ / _____ / _____ .

Carb/on/i/ferous
kär-bə-NIF-ə-rəs

50. The combining form -i/ferous means bearing or
producing. Build a word for the period in the earth's
history that produced many of the coal beds we have
today: _____ / _____ / _____ / _____ .

metall/i/ferous

51. Build a word that means bearing or producing
metal: _____ / _____ / _____ .

nickel/i/ferous

52. Build a word that means bearing or producing
nickel: _____ / _____ / _____ .

fluvi/o
FLŌŌ-vē-ə-graf

53. The combining form for river or stream is fluvi/o.
An instrument for measuring the rise and fall of a
river is called a _____ / _____ /graph.

fluvi/o/logy

54. Bio/logy is the science of dealing with life. What is a word for a science dealing with watercourses? _____/___/_____ .

fluvi/o/glacial

55. The verb erode comes from the Latin *erodere*, to eat away. There are several ways in which land can be eroded away: by wind, by sea, by stream, by lake, by volcanic action, or by glacial activity. When rock formations are formed by stream and volcanic action, they are fluvi/o/volcanic formations. When formations are created by stream and glacial action they are called _____/___/_____ formations.

fluvi/o/marine

56. The word used for sea is marine (Latin: *marinus*, sea). Build a word that means formed by the joint action of river and sea: _____/___/_____ .

fluvi/o/lacustrine
floo-vē-ə-LAK-ə-strən

57. The word used for lake is lacustrine (Latin: *lacus*, lake). Build a word that means formed by the joint action of lake and river: _____/___/_____ .

fluvi/o/aeolian
floo-vē-ō-ē-ŌL-ē-ən

58. The word used for wind is aeolian, from the Latin god of the winds, *Aeolus*. Build a word that means formed by the combined action of wind and river: _____/___/_____ .

soils

59. The combining form for soil is ped/o. Ped/o/logy is the study of _____ .

ped/o/genesis

60. If the word genesis means formation and development of, build a word that means formation of soil: _____/___/_____ .

61. By comparison with the word ge/o/sphere, build a word that means the soil layer of the earth.

ped/o/sphere

_____ / / _____ .

62. Look up the word root for mud. Write it here:

pel/o

_____ / __ .

63. Now, build a word that literally means mud stone: _____ / _____ .

pel/ite

64. Write down the combining form that means old or ancient: _____ / _____ .

pale/o

a discipline that deals with organisms (fossils) of the ancient past

65. Pale/o is often used in context of things or events that took place in the ancient past. What would be a good definition for the discipline of pale/o/biology? _____ .

deals with fossil plants of the ancient past

66. What about pale/o/botany?

_____ .

67. Lith is the word root for stone. What would be the word for ancient stone (a chipped stone used by ancient man)?_____ / / _____ .

pale/o/lith

68. The suffix -ic means of or relating to. Build a word that means relating to the Stone Age (of ancient stone [tools]): _____ / / _____ / ___ .

pale/o/lith/ic

69. What is a word that means, literally, foreign stone (a rock that is found included in another): _____ / / _____ . (Hint: What is the combining form for strange or foreign?)

xen/o/lith

hydr/o

70. Besides aqu/a, what is another combining form for water? _____/_____ .

hydr/o/lith

71. Stone that has dropped out of solution with water (like rock salt) is called _____/___/_____ .

aut/o

72. The combining form that means same is _____/_____ .

aut/o/lith

73. Stone that has been thrust into rock of a similar (same) kind is called an _____/___/_____ .

fire

74. When something is ignited, it is set on fire. The Latin word for fire, *ignis*, is the basis for the word igneous. An igneous rock is one that was formed by _____ .

igneous rock
IG-nē-əs

75. Volcanos spew out magma or molten rock. Another term for molten rock or rock created by fire is _____ _____ .

igneous breccia

76. Breccia (BRECH-ə) is a term for a rock consisting of fragments of rock embedded in a fine-grained matrix of sand or clay (like hard rice pudding). A breccia composed of rock formed from magma is called (two words) i_____ b_____ .

pseud/o/breccia

77. A rock that gives the *false* appearance that it is a breccia would be called _____/___/_____ . (Recall that the combining form for false is pseud/o.)

pseud/o/fossil

78. A natural object that looks like a fossil but is not is a _____ / ___ / _____ .

pyr/o

79. Another combining form for fire (the other is what?) is pyr/o. A pyr/o/maniac is a person who has an irrational impulse to start fires. That fire is involved is indicated by the combining form _____ / ___ .

volcanic magma
or very high
temperatures

80. In geology, pyr/o is often associated with volcanic magma or with very high temperatures. A pyr/o/meter is an instrument that measures _____ .

pyr/o/clast

81. The word root clast comes from the Greek word for rock fragment, *klastos* (broken). A fragment of rock ejected during a volcanic explosion (and very hot) is called a _____ / ___ / _____ .

clast/ic

82. When the suffix -ic is added to the word clast, it becomes _____ / ___ .

made of fragments
ejected from a
volcano

83. Clastic, an adjective, means made up of fragments of preexisting rocks. A pyr/o/clastic rock would be one _____ .

silic/i/clast/ic

84. Build a word that means made up of fragments of preexisting silica rock:
_____ /i/ _____ / ___ .

a rock made up of
the broken remains
of organisms

85. What do you think a bio/clast/ic rock is?
_____ .

seism/o

86. Write the combining form for earthquake here:
_____ / ___ .

seism/o/logy

87. If ge/ology is the science that deals with the earth, what is the name for the science that deals with earthquakes?
_____ / / _____ .

seism/o/meter or
seism/o/graph and
seism/o/logist

88. Predictably, the root seism comes from the Greek word for earthquake, _seismos_. Using seism/o, build a word for a device that measures earthquakes and a word for a scientist who studies them:
_____ / / _____ ;
_____ / / _____ .

seism/ic wave

89. A wave produced by an earthquake is a (two words) _____ / _____ _____ .

meta/morph

90. In ancient time, people believed that objects could be changed from one form into another by supernatural means. The Greeks called it meta/morph/o/sis (meta, change; morph, form; met-ə-MOR-fə-sis). Geologists call the process by which pressure, heat, and water change the pronounced change in a rock
_____ / _____ /ism.

meta/morph/ic

91. Build a word that means of or relating to metamorph/ism. (Hint: Remember the suffix for of or relating to.)
_____ / _____ / _____ .

metamorphic

92. What is a rock called that has gone through profound change during earth's history?
_____ rock.

igneous

93. So far, you have learned the names for two types of rocks. The rocks formed by volcanic action are called _____ rock. (Hint: fire.)

94. The rocks formed by the alteration of their character by pressure are called _____ rocks.

metamorphic

95. The third type of rock is formed by sediment (Latin: *sedimentum*, settling) where material is deposited by wind or water. When dirt settles to the bottom of a bucket of water, it is called _____ .

sediment

96. A sediment/ary rock is one that was created by _____ .

sediment

97. Shale was created from sediment, which would make it a _____ rock.

sedimentary

98. Sometimes, in metamorphism, a chemical is removed or added to the forming rock. The combining form, all/o means other. The word all/o/chemical would mean _____ .

other chemical

99. From the above information, form the term that would indicate that other chemicals were missing during metamorphism: _____/_____/_____ _____ .

all/o/chemical
metamorphism

100. The Greeks used the word *chthonios* to mean a spirit dwelling under the earth. Geologists use the term chthon/ic (THÄN-ik) to indicate a rock's dwelling. Build a word that means other dwelling (a rock that has been moved from its place of origin): _____/_____/_____ (drop the -ic).

all/o/chthon

101. What would be the term for a rock formation that remains at the *same* site of origin?

aut/o/chthon   _____/_____/_____ .

102. The word root trop means turn or change. The adjective combining form tropic is used in geology to mean tending to change. Substances that tend to change from one form or another are allotropic substances. Carbon can exist as either diamond or graphite. Carbon is an

allotropic   _____ substance.

103. Anis/o means unequal. The disease aniseikonia (an-ī-sī-KŌ-nē-ə) is an ailment in which the images of an object on each retina differ in size. Crystals of the same mineral tend to disperse light in different directions. The crystals, then, would be

anis/o/tropic   _____/___/_____ .

104. The combining form that means cold or freezing is cry/o. Cry/o/logy is the study of snow and ice, materials that exist at low temperatures. The study of glaciers would come under the field of

cry/o/logy   _____/___/_____ .

105. Would the mineral cry/o/lite most likely be found in a cold or warm region of the earth?

cold (Greenland)   _____ .

106. You have learned about the geosphere and the siderosphere. What is the term for the region of the earth's surface that is perennially frozen?

cry/o/sphere   _____/____/_____ .

107. What would be the term for the study of frozen soil? _____/__/_____/__/_____ (the study

cry/o/ped/o/logy   of permafrost).

strat/i

108. Write down the combining form that means layer: _____ / _____ .

stratum

109. Strat/i comes from the word stratum (STRA-dəm). In geology, stratum means a layer or bed of sedimentary rock. A bed of shale would be called a _____ of shale.

strata

110. The plural of stratum is strata, layers. Deposits of shale rock are often laid down in _____ .

strat/i/fied

111. To arrange in strata, or to lay down layers, is to strat/i/fy. Rock that has been laid down in strata has been (past tense): _____ / _____ / _____ .

stratified

112. Fluviomarine deposits are often laid down in layers of mud and sand. When such deposits become rock, the rock is called _____ rock.

strat/i/graph/y

113. Strat/i/graph/y is the science of rock strata. The study of stratified rock is the science of _____ / _____ / _____ .

strat/i/graph/ic

114. Build a word that means of or relating to stratigraphy: _____ / _____ / _____ .

strat/i/graph/er

115. What would be the term for someone who studies stratigraphy? _____ / _____ / _____ .

strat/i/graph/ic

116. What would a map be called that showed the strata of a particular area? A _____ / _____ / _____ map.

117. The Latin word for cave is *speleum*. Spele/o/logy (spē-lē-ÄL-ə-jē) is the study of caves. What would one call a person who specialized in this study?

spele/ol/o/gist   _____/___/___/_____ .

118. The Greek word for deposit is *thema*, from which comes the noun theme (thēm). Build a word that means a mineral deposit formed in a cave (a cave deposit). _____/___/_____

spele/o/them
(SPĒ-lē-ō-them)   (drop the final e).

   In this unit you worked with more than 70 new geological terms. Most of them are listed here for you to practice your pronunciation. Do that now. Then take the Unit Self-Test on the following page.

allochemical	fluviolacustrine	pelite
allochthon	fluviology	petrified
allotropic	fluviomarine	petrograph
anisotropic	fluviovolcanic	petroleum
aquifer	geochemistry	petrology
aquitard	geode	pseudobreccia
autolith	geodynamics	pseudofossil
bioclastic	geologist	pyroclast
calcarenite	geology	pyrometer
calcification	geomagnetic	sedimentary
calcite	hyalite	siderite
carbonaceous	hyalobasalt	siderosphere
carboniferous	hyalocrystalline	siliceous
cryology	hydrolith	siliciclastic
cryopedology	igneous	silicification
cryosphere	mineralogy	silicified
crystallite	metalliferous	speleology
crystallography	metamorphic	speleothem
cuprite	nickeliferous	stratified
ferriferous	paleobiology	stratigraphy
ferrimagnet	paleobotany	uraninite
fluvioaeolian	paleolith	xenolith
fluvioglacial	pedogenesis	
fluviograph	pedosphere	

Unit 5 Self-Test

TEST 1

From the list on the right, select the correct meaning for each of the following commonly used geological terms:

____ 1. fluviomarine
____ 2. igneous
____ 3. pseudofossil
____ 4. seismologist
____ 5. petroleum
____ 6. siderosphere
____ 7. speleothem
____ 8. pedology
____ 9. bioclast
____ 10. strata
____ 11. autochthon
____ 12. pelite
____ 13. metamorphic
____ 14. carboniferous
____ 15. xenolith

a. rock formation on the same site
b. rock made from broken
  remains of life
c. study of soils
d. mud stone
e. created coal beds
f. volcanic rock
g. layers
h. foreign stone
i. rock altered by pressure
j. by river and sea
k. rock oil
l. cave deposit
m. false organic remains
n. one who studies quakes
o. earth's iron core

TEST 2

Write the geological term for each of the following:

1. study of dynamic forces of earth _____
2. pertaining to earth's magnetic field _____
3. rock of calcium carbonate _____
4. clear and smooth as glass _____
5. study of crystals _____
6. rock that bears water _____
7. ore of copper _____
8. study of minerals _____
9. ferrous iron-containing mineral _____
10. science of rocks _____
11. mineral-replaced wood _____
12. bearing metal _____

13. study of rivers   _____
14. deals with ancient organisms   _____
15. rock fragment from volcano   _____

ANSWERS

Part 1

1. j	9. b
2. f	10. g
3. m	11. a
4. n	12. d
5. k	13. i
6. o	14. e
7. l	15. h
8. c	

Part 2

1. geodynamics	9. siderite
2. geomagnetic	10. petrology
3. calcite	11. petrified wood
4. hyaline	12. metalliferous
5. crystallography	13. fluviology
6. aquifer	14. paleobiology
7. cuprite	15. pyroclast
8. mineralogy	

UNIT 6

Organic Chemistry

In this unit you will build more than 60 organic chemistry terms. You will use some of the terms and parts you have already covered and also the following new prefixes, word-root combining forms, and suffixes.

meta- *(separated by one carbon atom on benzene)*
ortho- *(two groups on adjacent atoms)*
para- *(two groups on 1, 4 positions on benzene)*

amin/o *(–NH₂ group)*
brom/o *(bromine)*
but *(four carbon atoms)*
chlor/o *(chlorine)*
cycl/o *(ring)*
eth *(two carbon atoms)*
fluor/o *(fluorine)*
hydrox/y *(hydroxyl group [–OH])*
iod/o *(iodine)*

meth *(one carbon atom)*
nitr/o *(nitrogen)*
phenyl *(benzene with one hydrogen substitution)*
prop *(three carbon atoms)*
-al *(aldehyde)*
-amine *(–NH₂ group)*
-ane *(saturated hydrocarbon)*
-ate *(carboxylic salt)*
-ene *(double-bond hydrocarbon)*
-ine *(–NH₂ group)*
-oic *(carboxylic acid)*
-ol *(alcohol)*
-one *(ketone)*
-yl *(branch group)*
-yne *(triple-bond hydrocarbon)*

The International Union of Pure and Applied Chemistry (IUPAC) has developed a system for naming all organic compounds. By following the rules, a name can be assigned to a given structure, or the correct structure can be written from the name.

Use the following stem names to work this unit:

STEM NAME	# OF CARBONS	STEM NAME	# OF CARBONS
meth	1	hept	7
eth	2	oct	8
prop	3	non	9
but	4	dec	10
pent	5	undec	11
hex	6	dodec	12

These symbols will be used in this unit to represent atoms:

SYMBOL	ELEMENT	SYMBOL	ELEMENT
C	carbon	O	oxygen
H	hydrogen	I	iodine
N	nitrogen	Br	bromine
Cl	chloride	F	fluorine

carbon

1. The simplest class of organic compounds consists of the hydro/carbons. From the name, hydro/carbons contain two major types of atoms, hydrogen and _____ .

alk/ane

2. The suffix -ane corresponds to compounds known as alk/anes. Alkanes are saturated hydrocarbons; that is, each molecule contains the highest hydrogen-to-carbon ratio possible. Meth/ane is a hydrocarbon. It is also an _____/_____ .

alk/ane

3. Meth/ane has the formula CH_4. This hydrocarbon molecule is saturated; it contains the highest possible number of hydrogens-to-carbons. That makes this molecule an _____/_____ .

4. Meth/ane contains one carbon atom; prop/ane contains three carbon atoms. What is the name of an alkane that contains 6 carbon atoms?

hex/ane   _____/_____ .

5. Build a word that means an alkane that contains

do/dec/ane   12 carbon atoms: _____/_____/_____ .

6. In general, then, saturated hydrocarbons contain

-ane   the suffix _ _____ .

7. A but/ane carbon skeleton (with the hydrogen atoms removed for simplicity) can be drawn this way:

$$C—C—C—C \quad \text{(four carbon atoms)}$$

Draw a pent/ane skeleton:

C—C—C—C—C   _____ .

8. Alkanes can have branches of molecules attached to the main skeleton. When a branch is attached, the carbon atoms in the chain are numbered consecutively from that end of the chain nearest the branch. For example, a carbon branch for butane can appear in these ways:

$$\overset{\displaystyle X}{\underset{|}{C}}—\overset{}{C}—\overset{}{C}—C \quad or \quad C—\overset{\displaystyle X}{\underset{|}{C}}—C—C$$

$$or \quad C—\overset{\displaystyle X}{\underset{|}{C}}—C—C \quad or \quad C—C—C—\overset{\displaystyle X}{\underset{|}{C}}$$

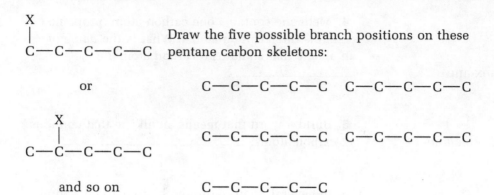

X
|
C—C—C—C—C

or

X
|
C—C—C—C—C

and so on

Draw the five possible branch positions on these pentane carbon skeletons:

C—C—C—C—C   C—C—C—C—C

C—C—C—C—C   C—C—C—C—C

C—C—C—C—C

9. The carbon atoms are numbered consecutively from that end of the chain *nearest* a branch or functional group. The following branched butane carbon skeletons would be numbered as shown:

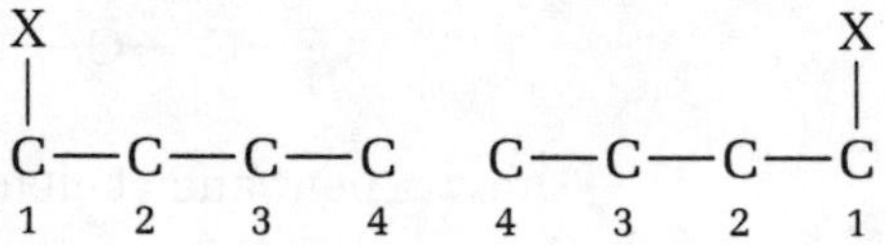

```
X                         X
|                         |
C—C—C—C       C—C—C—C
1   2   3   4       4   3   2   1
```

Number the carbon atoms in this branched pent/ane molecule:

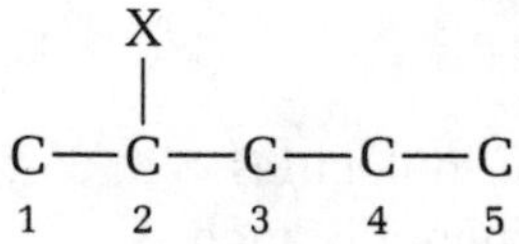

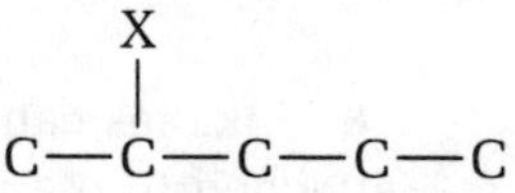

```
X
|
C—C—C—C—C
```

```
X
|
C—C—C—C—C
1   2   3   4   5
```

10. Each branch is located by the number of the atom to which it is attached on the main chain. Let's assume that a branch or functional group was called X on the following butane carbon skeleton:

```
X
|
C—C—C—C
1   2   3   4
```

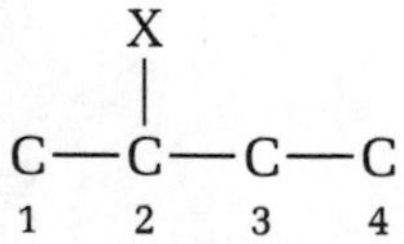

The branch would be called 2-X. Name the branch of this hypothetical pent/ane:

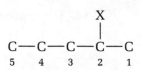

$$C—C—C—\overset{\overset{\displaystyle X}{|}}{C}—C$$
$$5\quad4\quad3\quad2\quad1$$

2-X
_____ .

11. Hydrocarbon branch groups are called alk/yl groups, and are named by the stem name plus the suffix -yl. For example, the branch of this butane:

$$C—C—\overset{\overset{\displaystyle CH_3}{|}}{C}—C$$

would be called 2-meth/yl (one carbon atom on position 2 of the carbon main chain). The

alk/yl   2-methyl group is an _____ / ____ group.

12. The entire molecule above would be named 2-meth/yl/but/ane (for a methyl group on the 2 position of the butane chain). The name of the branch group always precedes the name of the main chain. What would the name be for this molecule?

$$C—C—\overset{\overset{\displaystyle C}{|}}{C}—C—C—C$$

_____ -_____ / __ / _____ / _____

3-meth/yl/hex/ane   (Remember to number the carbon skeleton.)

13. What would be the name for this molecule? (Carbon skeleton shown only.)

$$C—C—C—\overset{\overset{\displaystyle C—C}{|}}{C}—C—C \quad \text{(careful)}$$

3-eth/yl/hept/ane   _____ -_____ / __ / _____ / _____

14. Try another.

$$
\begin{array}{c}
\qquad\qquad\qquad C—C—C \quad \text{(how many carbons?)}\\
\qquad\qquad\qquad\quad |\\
C—C—C—C—C—C—C—C \quad \text{(what branch}\\
\qquad\qquad\qquad\qquad\qquad\qquad\qquad\text{number?)}
\end{array}
$$

4-prop/yl/oct/ane   _____ -_____/ /_____ .

15. Again.

$$
\begin{array}{c}
\qquad\qquad\qquad\qquad\qquad\qquad C—C\\
\qquad\qquad\qquad\qquad\qquad\qquad\quad |\\
C—C—C—C—C—C—C—C—C—C
\end{array}
$$

3-eth/yl/dec/ane   _____ -_____/ /_____/_____ .

When two or more of the *same* type of alkyl group (branch) occur on a main chain, the group name is preceded by any one of the following prefixes or combining forms:

COMBINING FORM	MEANING	COMBINING FORM	MEANING
di	2	hept/a	7
tri	3	oct/a	8
tetr/a	4	non/a	9
pent/a	5	dec/a	10
hex/a	6		

The location of each alkyl branch on the main chain is indicated by a number. The numbers designating the positions of the alkyl groups are placed immediately before the names of the groups, and hyphens are placed before and after the numbers. If two or more numbers occur together, commas are placed between them.

16. This molecule:

```
                C
                |
    1    2      |    4
    C—C—C—C
         |
         C
```

is called 2,3-di/meth/yl/but/ane. The two methyl groups are on positions 2 and 3 of the main chain butane. The di- indicates that there are two methyl groups. What is the branch name for this molecule (include the position numbers, lowest number first)?

```
                  C
                  |
    1    2    3   |    5
    C—C—C—C—C
         |
         C
```

2,4-di/meth/yl   __ __ , ____ - ____ / _____ / ____ .

2,4-di/meth/yl/   **17.** What is the above alkane's name? _____ ,
pent/ane   _____ - _____ / _____ / _____ / _____ .

18. Try another one. What is the name for this hydrocarbon?

```
                        C—C
                        |
    1                   |
    C—C—C—C—C—C—C—C—C
         |         |
         C—C   C—C
```

3,5,6-tri/eth/yl/   _____ , _____ , _____ -
non/ane   _____ / _____ / _____ / _____ .

19. If several types of alkyl groups are present, they are placed in alphabetical order, and prefixed on the name of the basic alkane. The whole is written as a single word. For example, a molecule might be called 3-ethyl-2-methyl/pentane but never 2-methyl-3-ethyl pentane, because the letter *e* comes before the letter *m* in the alphabet. Which of these two molecular names is a correct name?

(a) 3-*e*thyl-5-*b*utyldecane

(b)

(b) 5-butyl-2,5,7,8-tetramethyldecane

20. Name this structure:

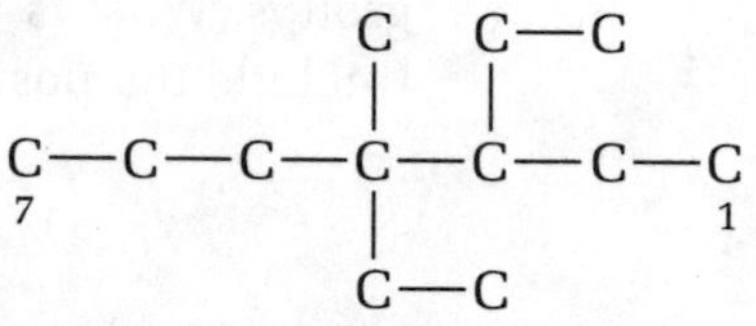

```
                    C    C—C
                    |    |
        C—C—C—C—C—C—C
        7           |        1
                    C—C
```

3,4-di/eth/yl-
4-meth/yl/hept/

3,_____ -di/_____ /yl/-4-_____ /yl/_____ /ane.

21. Try this one:

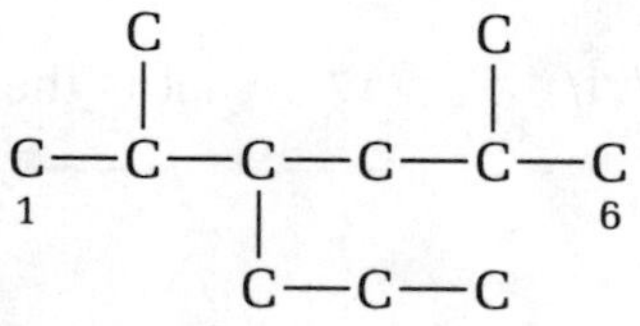

```
            C              C
            |              |
        C—C—C—C—C—C
        1       |          6
                C—C—C
```

2,5-dimethyl-
3-propylhexane

_____ .

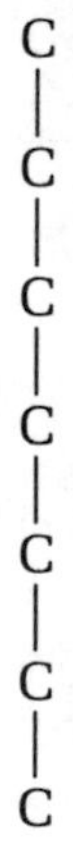

```
C
|
C
|
C
|
C
|
C
|
C
|
C
```

22. Draw a heptane carbon skeleton:

23. Place two ethyl groups at carbon atom 3.

```
      C                        C 1
      |                        |
  C—C   C                      C
      |   |                    |
C—C—C—C                        C
      |                        |
  C—C—C                        C
      |   |                    |
  C—C   C                      C
      |   |                    |
      C   C                    C
      |                        |
      C                        C
```

3,3-diethyl-2,4,
5-trimethyl-4-
propylheptane

Now, place a propyl group at carbon 4. Then, place one methyl group each on carbon atoms 2, 4, and 5. What is the name of the alkane?

_____ .

24. Many alkanes are arranged in a ring. For example, there is a pentane arranged this way:

```
        C
      /   \
    C       C
    \       /
     C — C
```

This structure is called cycl/o/pentane. In this case, the ringed structure of the molecule is designated by

cycl/o

the combining form _____/____ .

25. Build a word meaning ringed hexane.

cycl/o/hexane

_____/___/_____ .

26. In general, a ringed alkane is called a

cycl/o/alkane

_____/___/_____ .

double

27. When hydrocarbons contain double bonds, they are called alk/ene/s and are given the suffix -ene. But/ane contains only single bonds between the carbon atoms. But but/ene contains a _____ bond.

-ene

28. You can tell that but/ene contains a double bond in its structure because of the suffix _____ .

29. The root name for an alkene is derived from the longest continuous chain of carbon atoms containing the double bond. For example, this structure (only the carbon skeleton is shown):

$$C—C—C{=}C—C$$

is called pent/ene. What is the name for this structure?

$$C—C{=}C—C—C—C—C—C$$

oct/ene

_____/_____ .

-pent/ene

30. When an alkene contains two double bonds in its structure it takes the suffix -di/ene; if there are three, -tri/ene, and so forth. What is the suffix for an alkene with five double bonds? _____/_____ .

31. Numbering of the root chain is such that the carbons of the double bond have the lowest possible number. This chain:

$$C{=}C—C—C$$

is numbered this way:

$$\underset{1\quad\ 2\quad\ 3\quad\ 4}{C{=}C—C—C}$$

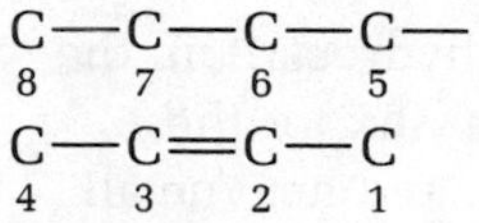

C—C—C—C—
8   7   6   5
C—C=C—C
4   3   2   1

Number the carbons on this root chain:

$$C—C—C—C—C—C=C—C$$

32. The position of the double bond is indicated in the name by placing the number of the lower-numbered carbon atom before the root name. Thus:

$$C=C—C—C$$

is called 1-but/ene. What is this alkene called?

$$C—C=C—C—C$$

2-pent/ene

_____/__ .

33. Remember that side chains or functional groups always come first in a name. This compound:

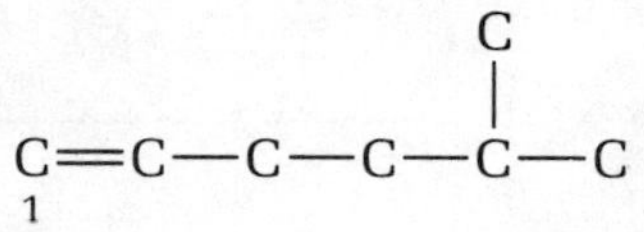

$$\begin{matrix} & & & & C & \\ & & & & | & \\ C=C—C—C—C—C \\ 1 & & & & & \end{matrix}$$

is called 5-methyl-1-hexene. Build a word for this alkene:

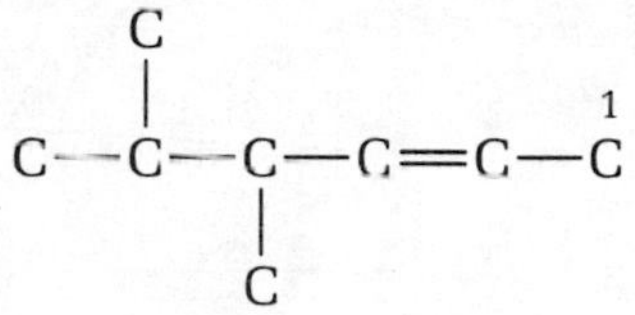

4,5-dimethyl-2-
hexene

$$\begin{matrix} & & C & & & & 1 \\ & & | & & & & \\ C—C—C—C=C—C \\ & & | & & & & \\ & & C & & & & \end{matrix}$$

34. Try this one. (Careful.)

$$C=C—C—C=C—C$$

1,4-hexa/di

1,_____ -_____/___ /ene.

-yne

35. When triple bonds appear in a hydrocarbon, the molecule is called an alk/yne. The suffix for these molecules is -yne. Butyne, pentyne, and hexyne all contain the suffix _____ .

but/yne

36. Build a word that means a hydrocarbon molecule with four carbon atoms and a triple bond:
_____ / _____ .

37. Name this chemical. Remember to number your carbon atoms:

$$C-C-C\equiv C-C$$

2-pent/yne

_____ - _____ / _____ .

38. Name this one, too:

$$C-C-C\equiv C-C-C$$

3-hex/yne

_____ - _____ / _____ .

39. Try this one (remember that the side groups come first):

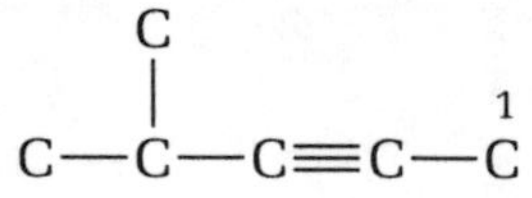

$$\begin{array}{c} C \\ | \\ C-C-C\equiv C-\overset{1}{C} \end{array}$$

4-methyl-2-pentyne

_____ .

-ane
single

40. Let's review quickly before we go on. Alkanes use the suffix _____ and contain _____ bonds.

-ene
double

41. Alkenes use the suffix _____ and contain _____ bonds.

-yne
triple

42. Alkynes use the suffix _____ and contain
_____ bonds.

43. Hydrocarbons also include the alcohols. Alcohols
contain hydroxyl groups (–OH). They are named
according to the longest chain that contains a
hydroxyl group. The ending -ol is substituted for
the *e* of the alkan*e*. Thus, butan*e* becomes butan/ol.
Change these alkanes into alcohols:

ethanol
propanol
pentanol
octanol
methanol

ethane _____ ;
propane _____ ;
pentane _____ ;
octane _____ ;
methane _____ .

44. Build a word for an alcohol with a chain of 10
carbon atoms: _____ .

decanol

45. The position of the hydroxyl group (OH) is
indicated by the number of the carbon atom to
which it is attached, with the numbering arranged
so that the hydroxyl group *receives the lowest
possible number.* For instance, the name of this
molecule

$$
\begin{array}{c}
\text{OH} \\
| \\
\text{C—C—C}
\end{array}
$$

is 2-propan/ol. The hydroxyl group is on the second
carbon atom. What is the name of this alcohol?

$$\text{HO—C—C—C—C}$$

1-butan

_____ - _____ /ol.

46. When alcohols contain two hydroxyl groups in their structures, their name includes the prefix di-, as in di/ol. When there are three hydroxyl groups, the prefix is tri-, as in tri/ol. This molecule:

$$\overset{\displaystyle OH}{\underset{\displaystyle \underset{3}{C}}{|}}$$

HO—$\underset{1}{C}$—$\underset{2}{C}$—$\underset{3}{C}$—$\underset{4}{C}$

is called 1,3-butan/di/ol. Build a word for this alcohol:

$$\overset{\displaystyle OH}{\underset{\displaystyle C}{|}} \qquad \overset{\displaystyle OH}{\underset{\displaystyle C}{|}}$$

C—C—C—C—$\underset{1}{C}$

$$\underset{\displaystyle OH}{|}$$

1,3,4-pentan/tri   _____ , _____ , _____ -
                        _____/_____ /ol.

47. Try this one:

$$\overset{\displaystyle C}{\underset{\displaystyle C}{|}}$$

C—C—C—C

$$\underset{\displaystyle OH}{|}$$

2-methyl-2-butan   _____ - _____ -2- _____ ol.

48. Again:

HO   CH$_3$
   \ C /
  C 1 C
  | |
  C   C
   \ C /
     4

1-methyl/cyclo/   _____ - _____ /cyclo/_____/_____ .
hexan/ol

49. The carbon-to-oxygen double-bonded group

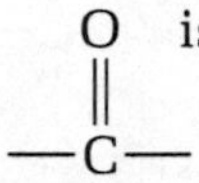

$$\underset{-\!\!\overset{\displaystyle\|}{\underset{\displaystyle}{C}}\!\!-}{O}$$ is called the carbonyl group.

When an alkyl (side) group and a hydrogen are attached to this group, the compound is called an aldehyde (AL-da-hīd). The IUPAC system uses the root name of the longest continuous chain containing the functional group and the ending -al for aldehydes. Chlorobutan/al is an

aldehyde   _____ . . .

50. . . . While chlorobutan/ol is an

alcohol   _____ . . . .

alkene   **51.** . . . And chlorobut/ene is an _____ .

52. The name of this aldehyde:

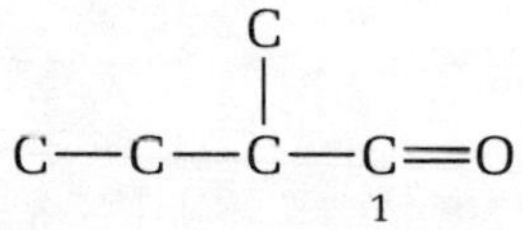

$$\underset{\qquad\quad\underset{1}{}}{C-C-\overset{\displaystyle C}{\overset{\displaystyle |}{C}}-C=O}$$

is 2-methyl/butan/al.
Build a word for this compound:

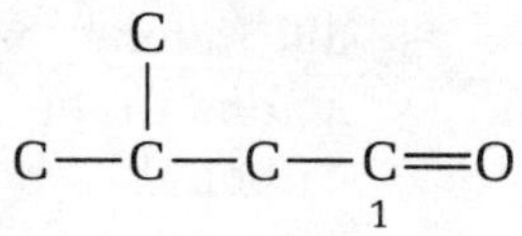

$$C-\overset{\displaystyle C}{\overset{\displaystyle |}{C}}-C-\underset{1}{C}=O$$

3-methyl/butan/al   _____ - _____ / _____ / ____ .

53. What is this aldehyde called?

$$C-C$$
$$|$$
$$C-C-C-C-C-C-C=O$$
$$|$$
$$C-C$$

3,5-di/ethyl/heptan   _____ , _____ -di/_____ / _____ /al.

54. When two alkyl groups are attached to the carbonyl carbon, the compound is called a ketone. The IUPAC system uses the root name for the longest continuous chain containing the functional group, and the ending -one, for ketones. The sex hormone testosterone (tes-TÄS-tə-rōn) is a _____ .

ketone

55. Try to build a word for this ketone:

$$O$$
$$||$$
$$C$$
$$C6 \quad ^1 \quad C$$
$$C \qquad 3C$$
$$C \qquad CH_3$$

3-methylcyclo-hexanone   _____ .

56. Ketones are also named by adding the word ketone to the names for the groups attached to the carbonyl carbon atom. A ketone with two ethyl groups is called di/ethyl ketone. What is the name of a ketone with two methyl groups?

di/methyl ketone   _____ / _____ _____ .

57. As a quick review, name the type of compound for each of these chemicals:

1. aldehyde
2. alkane
3. alcohol
4. ketone
5. aldehyde
6. alcohol
7. ketone

1. methylpropanal _____ ;
2. isobutane _____ ;
3. 1-propanol _____ ;
4. 3-phenyl-2-butanone _____ ;
5. ethanal _____ ;
6. 2-methyl-2-butanol _____ ;
7. 2-methyl/cyclo/hexan/one _____ .

58. List the number of carbon atoms in each of the main chains of the above molecules:

1. 3   5. 2
2. 4   6. 4
3. 3   7. 6
4. 4

1. _____   5. _____
2. _____   6. _____
3. _____   7. _____
4. _____

59. Organic compounds that contain the $-NH_2$ group in their main chains are called amines. Many amines are named by prefixing the names of the attached groups to the word amine. If identical $-NH_2$ groups are present, the prefixes di- and tri- are used. This molecule:

$$C—C—N$$

is called ethyl/amine. What is the name for this compound?

$$C—C—C—N$$

propyl/amine   _____/_____ .

60. Try this one. Remember that side groups are named first and go in alphabetical order.

$$
\begin{array}{c}
C \\
| \\
C—C—C—N \\
| \\
C—C
\end{array}
$$

_____ .

amine

61. While it is not universal, more complicated amines have the ending -ine in their names. Quinoline is an _____ .

amines
nitrogen

62. Two chemicals, purine and pyrimidine, make up a large part of the DNA molecule. Purine and pyrimidine (pi-RIM-ə-dēn) are _____ because they contain _____ in their main structure.

63. Build a word for this molecule:

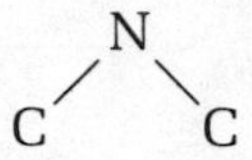

$$\begin{array}{c} N \\ \diagup \quad \diagdown \\ C \qquad C \end{array}$$

di/methyl

_____/_____ /amine.

64. Try this one:

$$N—C{=}C—N$$

ethyl/ene/di/amine   (Careful) _____/_____ /di/ _____ .

Use these combining forms for the following section.

NAME	FUNCTIONAL GROUP (SIDE CHAIN)
fluor/o	$-F$
chlor/o	$-Cl$
brom/o	$-Br$
iod/o	$-I$
nitr/o	$-NO_2$
amin/o	$-NH_2$
hydrox/y	$-OH$

65. In chemistry, the combining form chlor/o indicates that the molecule contains the element chlorine. Chlor/o usually precedes the rest of the molecular name. Chlor/o/benzene, chlor/o/form, and chlor/ambucil all contain the element

chlorine _____ in their structures.

66. The combining form chlor/o can also mean yellowish green. The material in plants that makes food from sunlight is called chlor/o/phyll (KLOR-ə-fil; Greek: *phyllon*, leaf). Chlorophyll has a

yellowish-green _____ color.

67. What would you suggest chlor/o means in the

Chlorine word di/chloro/ethylene?_____ .

The fragrant oils of clove and wintergreen belong to a class of compounds called aromatic, named because of their odor. The simplest member of this class is benzene (benz/ene; BEN-zēn), a cyclic, six-carbon molecule containing one hydrogen attached to each carbon atom:

68. The benzene ring with one hydrogen removed is called a phenyl group (FEN-l). For compounds in which a benzene ring is attached to a larger group of carbon atoms, the term phenyl is used as a prefix in the same way as methyl. Phenyl/alanine is an essential amino acid. That its structure contains a benzene ring is indicated by the prefix

phenyl _____ .

phenyl

69. Build a word for a benzene-containing drug that raises blood pressure: _____ /ephrine.

70. In many cases, when a hydrogen of benzene is replaced by one or more simple atoms (bromine, nitrogen, chlorine, to name a few), the first part of the name indicates the replacing atom, and the second part is the word benzene. Brom/o/benzene contains the atom bromine substituted in the

benzene

compound _____ .

71. Build a word for a molecule containing a chlorine atom attached to a benzene ring:

chlor/o/benzene

_____ / ___ / _____ .

72. When two groups are substituted on benzene, their relative positions on the molecule are indicated with either of three prefixes, ortho- (abbreviated as o), meta- (abbreviated as m), or para- (abbreviated as p). Ortho- means that the two groups are on adjacent carbon atoms on the benzene ring. Meta- means that the substituted carbon atoms are separated by one carbon atom. Para- means that the two groups are in the 1,4 positions. O-, m-, and p-di/bromo/benzene look like this (be sure to note where the bromine atoms are located on the benzene ring):

o-dibromo   m-dibromo   p-dibromo

Now build a word for this molecule:

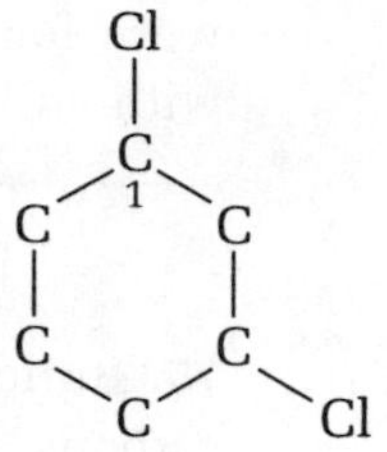

m-di/chlor/o/
benzene

_____ - _____ /chlor/o/ _____ .

73. When three or more substituted atoms are present on the benzene ring, a numbering system is used. In most cases, a carbon atom bearing a substitution is called number 1, and the ring is numbered so that the substituents are located on low-numbered atoms. This molecule:

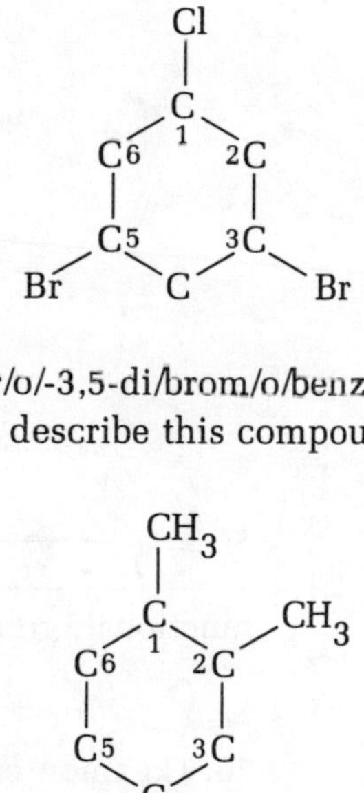

is called 1-chlor/o/-3,5-di/brom/o/benzene.
Build a word to describe this compound:

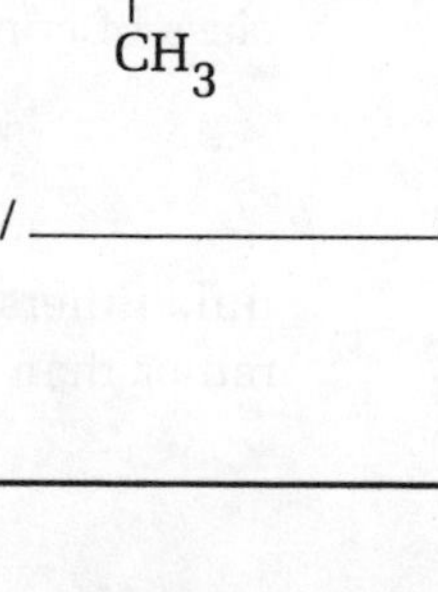

1,2,4-tri/methyl/
benzene

1,_____ ,_____ -tri/ _____ / _____ .

74. The combining form iod/o indicates that a compound includes the element iodine. Build a word for an alkane with a main chain of four carbons with an iodine atom attached:

iod/o/butane

_____/_____/_____ .

75. Build a word for an alcohol that contains eight carbons, with an iodine atom on the second carbon.

2-iod/o/octan/ol

_____ - _____/_____/_____/_____ .

76. The combining form fluor/o is used to indicate that a compound contains the element fluorine. Build a word for an alkene made of six carbon atoms and that contains a fluorine as a functional group:

fluor/o/hex/ene

_____/_____/_____/_____ .

77. Build a word for this benzene molecule:

1,3,4-tri/fluor/o-
2,6-di/iod/o/
benzene

1, _____ , _____ tri/_____/_____ -2, _____ -di/ _____/_____/_____ (Remember to place functional groups into alphabetical order.)

78. Common ether is a highly flammable liquid that is often used as an anesthetic. It belongs to a general class of compounds called ethers, which contain a

$$C—O—C$$

link. Ethers are named using the functional group rather than the longest chain as the parent.

The compound

$$CH_3 - O - CH_3 \qquad (C - O - C)$$

is called di/methyl ether. What is this ether called?

$$C - C - O - C - C$$

di/ethyl ether   _____/_____ _____ .

79. Build a word for an ether that contains two propyl groups:

di/propyl ether   _____/_____ _____ .

80. When the two substituents are different, the groups precede the word ether in alphabetical order. Build a word for an ether that contains a one-carbon group and a two-carbon group:

ethyl methyl ether   _____ _____ _____ .

Use the following chart to work the next section.

NAMES OF THE COMMON ACIDS		
IUPAC NAME	*COMMON NAME*	*DERIVATION*
methan/oic	formic	Latin: *formica*, ant
ethan/oic	acetic	Latin: *acetum*, vinegar
propan/oic	propionic	Greek: *protos pion*, first fat
butan/oic	butyric	Latin: *butyrum*, butter
pentan/oic	valeric	Latin: *valere*, powerful
dodecan/oic	lauric	laurel
hexadecan/oic	palmitic	palm
octadecan/oic	stearic	Greek: *stear*, tallow

81. Compounds that contain the carboxyl group

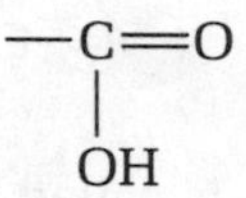

$$-C=O$$
$$|$$
$$OH$$

make up a large family of natural materials. For instance, acetic acid, which has been known since biblical times, is more commonly known as vinegar. Many other carboxylic acids are known by their common names. However, IUPAC nomenclature employs the suffix "-oic acid" added to the hydrocarbon stem minus its final *e*. For example, the name methan/e can be converted to methan/oic acid this way. Convert these alkanes to carboxylic acids:

propan/oic acid   propane _____/____ _____ ;
pentan/oic acid   pentane _____/____ _____ ;
dodecan/oic acid   dodecane _____/____ _____ ;
octadecan/oic acid   octadecane _____/____ _____ .

82. Build a word for a carboxylic acid that has a main chain of three carbon atoms with a chlorine functional group:

_____/ /_____/__ _____ .

chlor/o/propan/oic   (Remember that functional group names precede the
acid   name for the main chain.)

83. Try another. Build a word for a carboxylic acid having a main chain of four carbon atoms with an
brom/o/butan/oic   attached bromine atom:
acid
_____/ /_____/__ _____ .

84. Carboxylic acids react with base to produce salts of the acids. The names of salts are derived from the names of the acids by changing the -ic ending to -ate and preceding this name by the name of the metal involved in the chemical reaction. When sodium (Na) hydroxide (a strong base) reacts with ethano/ic acid, the product is sodium ethano/ate. Build a word for the product of a reaction with propanoic acid and a base containing calcium (Ca):

calcium
propano/ate   _____ _____/___ .

85. Try another. Build a word for the product of a reaction with undecanoic acid and a base containing zinc (Zn):

zinc undecano/ate   _____ _____/___ .

86. Try this: Change octane to a carboxylic acid and then give the product of that acid with a base containing potassium:

1. octanoic acid
2. potassium
   octanoate

1. _____ ;
2. _____ .

In this unit you formed more than 60 new words. Forty-eight of these terms are listed below to allow you to practice your pronunciation. Pronounce each term several times before you take the Unit Self-Test.

alcohol	cyclopentane	methylbutanol
aldehyde	dimethylbutane	methylcyclo-
alkane	dimethyl ketone	hexanone
alkene	ether	octadecanoic acid
alkyl	ethylamine	octanol
alkyne	ethylheptane	pentane
amine	ethyl methyl ether	pentanoic acid
benzene	ethylmethylpentane	pentene
bromobenzene	fluorohexane	phenylalanine
bromobutanoic acid	hexyne	phenylephrine
butandiol	hydrocarbon	purine
butane	iodobenzene	pyrimidine
butyloctane	ketone	quinoline
butyne	methane	testosterone
chlorobenzene	methanol	trimethylbenzene
chlorobutanal	methylbutane	undecanoate
cycloalkane		

Unit 6 Self-Test

PART 1

From the list on the right, 'select the correct category for each of the following chemicals:

___ 1. bromobutanoic	a. alkyne
___ 2. pyrimidine	b. alcohol
___ 3. octanol	c. ketone
___ 4. ethylheptane	d. benzene
___ 5. undecanoate	e. alkane
___ 6. methylbutanol	f. amine
___ 7. methylcyclohexanone	g. salt
___ 8. chlorobenzene	h. aldehyde
___ 9. hexyne	i. carboxylic acid
___ 10. phenylalanine	
___ 11. chlorobutanal	

PART 2

Build terms for the following structures:

1. C—C—C ______________________
2. C—C—C—C—C—C ______________________
3. C—C—C—C—OH ______________________
4. ______________________
5. C—C—C—C=C—C—C—C—C=C=C ______________________
6. C—C—N ______________________
7. ______________________
8. C≡C—C—C—C—C—C—C—C—C ______________________

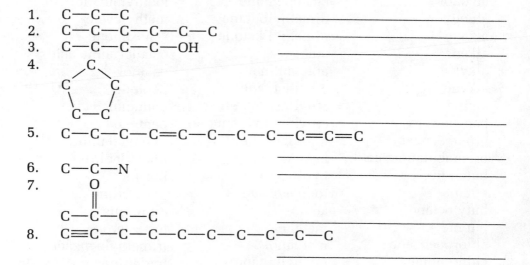

PART 3

Draw the structure from the following names (it is not necessary to include the hydrogens):

1.   6-chloro-5,8-diethyl-8-iodo-3,4,4-trimethyl-2,6-decadiene

2.   4-ethyl-3,4-dimethyl-2-hexene

ANSWERS

Part 1

1. i	7. c
2. f	8. d
3. b	9. a
4. e	10. d
5. g	11. h
6. b	

Part 2

1. propane
2. hexane
3. butanol
4. cyclopentane

5. 1,2,7-undecatriene
6. ethylamine
7. 2-butanone
8. 1-decyne

Part 3

1.

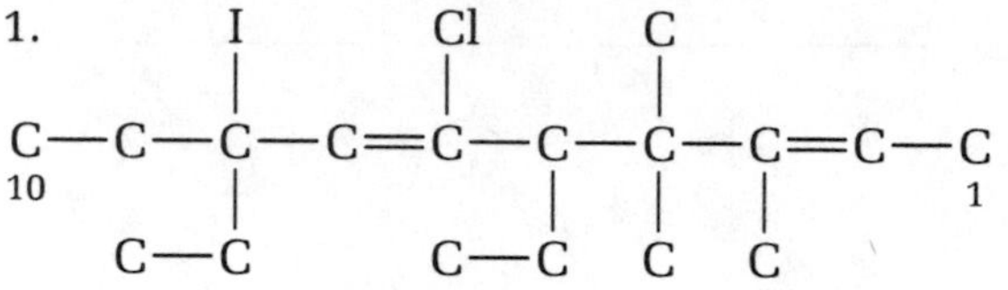

```
                I          Cl         C
                |          |          |
  C—C—C—C=C—C—C—C=C—C
  10      |          |    |    |           1
        C—C        C—C  C    C
```

2.

```
              C
              |
  C—C=C—C—C—C
  1    |    |        6
       C    C—C
```

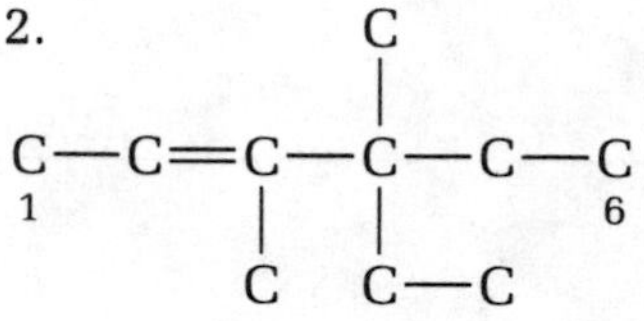

UNIT 7

Cell Biology

In this unit you will put together at least 50 new words, using the following prefixes, root combining forms, and suffixes.

ana- *(up, upward)*
anti- *(against)*
auto- *(self)*
cata- *(down, against)*
co- *(with, shared)*

adip/o *(fat)*
ax/o *(axon, axis)*
bacteri/o *(bacteria)*
centr/o, /i *(center)*
cyt/o *(cell)*
dendr/o *(like a tree)*
gamet/o *(gamete)*
gli/o *(glue)*
is/o *(equal)*
kary/o *(nucleus)*
kinet/o *(movement)*
lemm/o *(shell, husk)*
lys/o *(loosening)*

mit/o *(thread)*
nucle/o *(nucleus)*
olig/o *(few, small)*
o/o *(egg)*
phag/o *(eating)*
reticul/o *(net)*
sarc/o *(flesh)*
spermat/o *(sperm)*
trop/o *(turning)*
vir/o *(virus)*

-chore *(move apart)*
-oides *(like)*
-osis *(formation)*
-ote *(native)*
-plasm *(form)*
-plast *(formed)*
-some *(body)*

cells

1. The root combining form cyt/o means cell. Cyt/o/logy (sī-TĂL-ə-jē) is the study of _____ .

cells

2. A cyt/o/log/ist is one who studies _____ .

Please refer to the following diagram as you work in this unit.

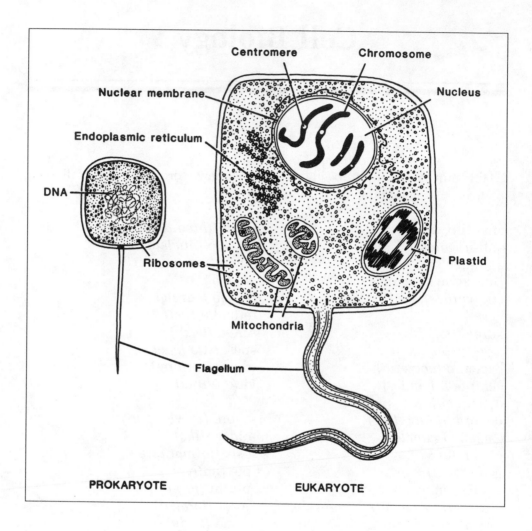

Centromere

Chromosome

Nuclear membrane

Nucleus

Endoplasmic reticulum

DNA

Ribosomes

Plastid

Mitochondria

Flagellum

PROKARYOTE   **EUKARYOTE**

a substance
poisonous to cells

3. Cyt/o/toxin uses the combining form cyt/o before the English word for poison. Cyt/o/toxin means
_____ .

cytotoxin

4. Cyanide is poisonous to cells. That makes cyanide a _____ .

cyt/o/some

5. The word-root combining form for body is -some (sōm). Form a term that means, literally, body of the cell, but that is taken to mean that portion of the cell material outside of the nucleus:

_____/___/_____ .

cyt/o/skeleton

6. A skeleton is a supporting framework. Build a word that means supporting framework of the cell.

_____/___/_____ .

cell

7. Cyte, at the end of a word, also means

_____ .

erythr/o/cyte

8. The combining form for white is leuk/o. A leuk/o/cyte is a white blood cell. Form a term that means red blood cell:

_____/___/_____ .

(Look in Unit 2, if you've forgotten.)

nucle

9. Most cells contain one or more bodies that contain the genetic material of the cell. Such a body is called a nucleus. Write the word root for nucleus:

_____ .

relating to
the nucleus

10. The suffix -ar makes a word an adjective, and means of or relating to. What does the adjective nucle/ar mean?

_____ .

nucle/ar

11. The membrane encompassing the nucleus would be the _____/___ membrane.

cyt

12. The combining form -plasm means form. The material that gives the cell form is called _____ /o/plasm.

nucle/o/plasm

13. What gives the nucleus form?

_____ /o/_____ .

nucle/o/protein

14. The general term for a protein that is found in the nucleus is a _____ / / _____ .

little nut

15. The term nucleus means nut in Latin. Within the nucleus is a smaller structure called the nucleolus. Nucleolus must mean _____ _____ in Latin.

nucleolus
nōō-KLĒ-ə-ləs

16. The root for nucleolus is nucleol. Nucleol/oid means resembling a _____ .

nucle/ate

17. The adjective suffix -ate means having. Chord/ate means having a structure called a notochord. Build a word meaning having a nucleus: _____/_____ .

nucleol/ate

18. Now build one to mean having a nucleolus: _____/_____ .

chrom

19. The nucleus harbors the genetic material of the cell, called chromatin. Chromatin is made of DNA and proteins. The word root in chrom/atin is _____ .

chrom/atin

20. Chrom means color (or colored). Special cellular dyes color chromatin, which is how it got its name. The material that is colored in the nucleus is _____/_____ .

chrom/o/some

21. At special times, the genetic material condenses into unique bodies of chromatin. Form a word that means colored body: _____ /o/_____ .

chrom/o/somes

22. The human body has 23 pairs of chrom/o/somes. But some animals have only a few _____/ /_____ .

eu/chrom/atin
yōō-KRŌ-mə-tin

23. Build a term that means true chromatin: _____/ /_____ .

euchromatin

24. When the cell is well and active, the cell's nucleus has much _____ .

nucleus

25. Another combining form that means nucleus is kary/o. Kary/o/logy (kair-ē-ÄL-ə-jē) is the science of the study of the cell's _____ .

kary/o/logist

26. What do you call the specialist who studies the nucleus? _____/ /_____ .

karyoplasm

27. What term means the same as nucleoplasm? _____ .

nucle/o
kary/o

28. Therefore, both the combining forms _____/ and __ _____/ mean nucleus.

eu/kary/ote

29. The suffix -ote makes a word a noun and means native. Using -ote and kary, build a word for an organism that is composed of cells, and in which each cell has a true nucleus (plural: nuclei): _____/ /_____ .

eukaryote

30. An ameoba is an organism with a true nucleus. That makes the ameoba a _____ .

eukaryotes

31. Humans also have cells with nuclei. Humans are also _____ .

natives before
nuclei

32. Bacteria do not have nuclei. Bacteria are pro/kary/otes. What is the literal definition of this term? _____
(See Unit 1, if you've forgotten pro-.)

prokaryotes
eukaryotes

33. Bacteria evolved before the first cells with nuclei came on the scene. Now, both types of organisms exist. Name them: _____ ;
_____ .

threadlike

34. The powerhouse of the cell is the mit/o/chondri/on (Latin: *chondrion*, small grain). From the combining form for this term, the structure must appear both grainlike and _____ .

mit/o/chondri/a

35. To make the plural of this term, you remove the -*on* and replace it with -a. Build a term that means more than one mitochondrion:
_____/ /_____/_____ .

mit/o/chrondri/al

36. The suffix -al makes a term into an adjective. Build a word that means relating to mitochondria:
_____/ /_____/_____ .

mit/osis

37. The combining term mit/o also is used in the name of the process of mit/osis (Latin: *mitos*, thread + *osis*, formation) or cell division. Thus, when amoeba divide in two they go through
_____/_____ .

mit/otic

38. Changing -osis to -otic makes the term into an adjective. Hypn/osis (sleep formation) becomes hypn/otic. Make mitosis an adjective:
_____/_____ .

mitotic

39. Chromosomes are _____ figures.

mitotic

40. The count of the number of actively dividing cells in a group of cells is called the _____ index.

sarc/o
sar-kə
like flesh

41. Look back in the list. The combining form for flesh is _____/_____ . Sarc/oid means

_____ .

sarc/o

42. Sarc/o is used when describing the structures of a muscle cell (flesh =muscle). The semifluid substance in muscle cells is called _____/____ /plasm.

sarc/o/plasm/ic

43. Using the suffix -ic, make sarcoplasm into an adjective: _____/___/_____/____ .

reticul
reticul

44. Look up the word root for net and write it here: _____ . The deep layer of skin is made up of a net of fibers and it is therefore called the _____ /ar layer.

sarc/o/plasm/ic
reticul/um

45. Reticul/um means network. The network of certain protein-producing tubes in the muscle cell is called the (two words) _____/___/_____/___

_____/____ .

formed within
network formed
within

46. End/o/plasm means _____ . The end/o/plasm/ic reticulum structure responsible for making proteins in cells means, literally,

_____ .

endoplasmic
reticulum

47. The genetic instructions from the nucleus are brought to the _____ _____ for processing.

sarcoplasmic
reticulum

48. In the muscle, instructions are brought to the
_____ _____ .

lemm

49. The word root meaning shell or sheath is
_____ .

neur/o
sarc/o/lemma

50. The delicate sheath surrounding a nerve cell is
called a _____/_____ /lemma. The sheath around a
muscle cell is called a _____/___/_____ .

fat
fat cell

51. Adip/o means _____ . An adip/o/cyte is a
_____ .

52. Adip/o/genesis means
_____ .

formation of fat
fatty white cell

Adip/o/leuk/o/cyte means
_____ .

Note the terms in this picture of a nerve cell:

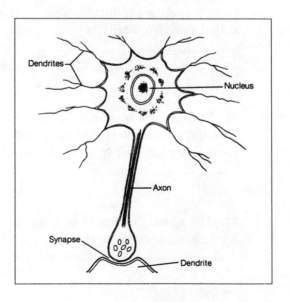

53. The combining form meaning axis is
_____/____ . The ax/on of a nerve cell is the
_____ , or long, slender process that
conducts signals from the cell body to another cell.

ax/o
axis

54. Ax/o/plasm is the fluidlike material found in the
_____ of nerves.

axons

55. Refer to the list. The word root that means like
a tree is _____ . Dendr/ites are nerve cell
processes that collect nerve signals from other cells;
the dendrites give the nerve cell the appearance of a
_____ .

dendr
tree

56. Analyze ax/o/dendr/ite:
_____/_____ (combining form);
_____ (word root);
_____ (suffix).
Build the word: _____/____/_____/_____ .
What do you think it means?
_____ .

ax/o
dendr
-ite
ax/o/dendr/ite
treelike native of
axon; dendrite that
branches from axon

57. The word-root combining form for few or small is
_____/____ .

olig/o

58. Analyze olig/o/dendr/o/cyt/e
_____/_____ (combining form);
_____/_____ (combining form);
___/___ (word root);
_____ (ending).
Build it here:
_____/___/_____/__/_____/_____
What does it mean?
_____ .
(Now look the word up in your dictionary.)

oliglo
dendr/o
cyt
-e
olig/o/dendr/o/cyt/e
cell having few
treelike branches

gli
glue

59. Glia is the Greek word meaning glue. The word root meaning glue is _____ . Neur/o/gli/a is the _____ or supporting tissue of nerves.

gli/o/cyte

60. Build a word that means a cell that makes glia: _____ / _____ / _____ .

eating cell

61. Phag/o/cyte means _____ .

phag/o/cyte
FAG-ə-sīt

62. A cell that engulfs foreign material within the body is generally called a _____ / _____ / _____ .

phag/o/cyt/osis
phagocytosis

63. Using the suffix -osis, build a word that means the process of ingestion (of foreign material) by cells: _____ / _____ / _____ / _____ . The bulk ingestion of solid material by cells is also called _____ .

process of drink-
ing; ingestion of
fluid by cells

64. The Greek word *pinein* means to drink. What does pino/cyt/osis mean? _____ .

process within the
cell

65. End/o/cyt/osis incorporates both events. What does the term mean? _____ .

lys/o

66. Lys/o/zyme is an enzyme found in saliva and tears that loosens or dissolves the coats of many bacteria. What is the combining form for this term? _____ / _____ .

lys/o/some
LIS-ə-sōm
harbors enzymes
that digest foreign
material

67. Build a word that means dissolving
body, for the saclike structures found in cells:
_____/_/_____ .
What do you think this structure does?
_____ .

phag/o/some
heter/o/phag/o/
some

68. Build a word that means:
eating body: _____/_/_____ ;
different eating body
___/_/___/_/_____ .

cyt/o/lys/o/some
auto/phag/o/some
harbors enzymes
used to digest cell

69. Build a word that means cell-dissolving body:
___/_/___/_/_____ .
Build one that means self-eating body:
_____/_____/_/_____ .
What does this cellular body do?
_____ .

vir/o
bacteri/o

70. Foreign material includes bacteria and viruses.
Look in the list. What are the combining forms for
these words?
virus _____/___ ;
bacteria _____/___ .

viruslike

71. Vir/oid means _____ .

vir/osis
vir/o/logy

72. Build a word that means:
virus disease _____/_____ ;
science that deals with viruses
_____/_/_____ .

73. Build a word that means:
produced by bacteria
_____/_/_____ /ic;
the science dealing with bacteria
_____/_/_____ ;

destruction of bacterial cells

bacteri/o/gen   _____/___/_____ /is;

bacteri/o/logy   bacterial disease

bacteri/o/lys   _____/_____ ;

bacteri/osis   stopping or checking bacterial growth

bacteri/o/stasis   _____/___/_____ .

74. O/o is the combining form for egg. An o/o/meter (ō-ÄM-ə-dər) is an instrument for measuring

eggs   _____ .

75. The egg is a single cell. Build a word that means: precursor egg cell

_____/___/_____ ;

egg cell

_____/___/_____ ;

formation of the egg

_____/___/_____ /esis;

o/o/blast   a membrane or sheath surrounding the egg

o/o/cyte   _____/___/_____ /a;

o/o/gen   instrument for viewing the inside of the egg

o/o/lemm   _____/___/_____ ;

o/o/scope   egg form (the fluidlike material in the egg)

o/o/plasm   _____/___/_____ .

76. Analyze o/o/spor/i/fer/ous:

_____/_____ (combining form);

o/o   _____/____ (combining form);

spor/i   _____ (word root);

fer   _____ (suffix).

-ous   Build it:

o/o/spor/i/fer/ous _____/___/___/___/___/___/_____ .

bearing egg spores What does it mean?
(plants)   _____ .

77. The combining form for sperm is

spermat/o   _____/____ .

78. Using the root for sperm, form a word meaning:
resembling sperm or a sperm cell
_____/_____ ;
loosening or dissolving sperm
_____/_____/_____ /is;

spermat/oid
spermat/o/lys
spermat/o/gen
spermat/o/plasm

the whole sperm formation
_____/_____/_____ /esis;
the plasm of a sperm cell
_____/____/_____ .

trop/o

79. Go back to the list. The word-root combining
form that means turning is _____/_____ .

80. Heli/o/trop/ism means the tendency
to turn towards or affinity for the sun.
What does neur/o/trop/ism (-trō-piz-əm) mean?

affinity for nerves   _____ .

trop/o/myosin

81. Using the combining form for turning, build a
word that means affinity for myosin, the protein of
muscle: _____/____/_____ .

trop/o/collagen

82. Now build a word that means affinity for
collagen, the protein of connective tissue:
_____/____/_____ .

gamet/o
gametes

83. Back to the list again. The combining form for
gamete is _____/_____ . Both sperm and egg cells
are _____ .

gamet/o/gen/esis
gə-mēd-ə-

84. The production of gametes is called
_____/____/_____/_____ .

is/o/chromosome

85. Is/o is the combining form that means same, uniform, or equal. Is/o/gon/ic means having equal angles. A chromosome with equal arms is called an _____/_____/_____ .

same size as
another gamete

86. Is/o/gamet/e means _____ .

a gamete that is a
different size
than the other
(sperm vs. egg)

87. What is a heter/o/gamet/e? _____ .

equal or all
directions

88. A cell that is is/o/trop/ic grows in _____ directions.

centr/o

89. The Greek word for part or segment is *meros*. Using the combining form for central, build a word that means the central part of a chromosome: _____/_____ /mere.

chrom/o/mere

90. Now build one that means colored segment, for the colored material in chromatin: _____/_/_____ .

acr/o

91. When the centromere is on the extremity or tip of the chromosome, the chromosome is termed _____/_____ /centr/ic.

centr/ic

92. When the centromere is in the center of the chromosome, it is called meta/_____/_____ .

93. The disk-shaped structure attached to
the centr/o/mer/ic chromatin is the kinet/o/chore.
Locate the word parts in this term:＿＿＿＿＿／＿＿ ;

kinet/o
-chore
movement apart

＿＿＿＿＿＿＿ .
What does the term mean?

＿＿＿＿＿＿＿＿＿＿＿＿ .

kinetochore
kə-NĒD-ə-kōr

94. The structure that plays a role in the movement
of chromosomes is the ＿＿＿＿＿＿＿ .

95. In plant cells, the suffix -plast often connotes
a saclike body filled with pigment. A chlor/o/plast is

green
colored

filled with ＿＿＿＿＿＿ pigment; a chrom/o/plast
is filled with ＿＿＿＿＿＿ pigment.

96. The activity that occurs within the cell is called
metabolism, from the Greek word for change,
metabole. Metabolism is actually two processes: one
in which food is broken down for energy, and one
in which food energy is used to break up cellular
structures and make new ones. With that in mind,
use the stem -bolism (Greek: *ballein*, to throw) and
the prefixes in the list at the beginning of the unit
to form a term that means:
"thrown down," destructive metabolism

＿＿＿＿＿＿＿／＿＿＿＿ ;

cata/bolism
ana/bolism

"thrown upward," constructive metabolism

＿＿＿＿＿＿＿／＿＿＿＿ .

97. Look at the list again. Write down the prefix that

anti-
co-

means against: ＿＿＿＿＿＿＿ . Now write down the
one that means with or shared: ＿＿＿＿ .

98. Build a word that means with an enzyme:

co/enzyme

＿＿＿＿／＿＿＿＿＿＿＿ .

99. Now build a word that means against a body:

anti/body   _____/_____ .

100. Histamine is a chemical produced by mast cells that causes runny noses and red, ichy eyes. Build a term for the drug used against this chemical:

anti   _____ /histamine.

In this unit you worked with more than 100 terms of cell biology. Thirty of them are listed here for you to practice your pronunciation. Do that first, and then take the Unit Self-Test.

acrocentric	euchromatin	neurolemma
antibody	eukaryote	neurotropism
axodendrite	gametogenesis	oligodendrocyte
bacteriolysis	heterogamete	oolemma
catabolism	karyoplasm	ooplasm
centromere	kinetochore	phagocyte
chromoplast	lysosome	sarcoplasmic
coenzyme	metacentric	spermatogenesis
cytolysosome	mitochondria	tropocollagen
endocytosis	mitosis	virosis

Unit 7 Self-Test

PART 1

From the list on the right, select the correct meaning for each of the following often-used cell biology terms.

___ 1. cytosome	a. when cell is active
___ 2. nucleoplasm	b. network
___ 3. chromosome	c. eating cell
___ 4. euchromatin	d. sheath around egg
___ 5. eukaryote	e. process of drinking by cell
___ 6. reticulum	f. affinity for nerves
___ 7. nucleolus	g. different-sized sex cell
___ 8. adipocyte	h. central part
___ 9. oligodendrocyte	i. cell body
___ 10. phagocyte	j. tip of chromatin
___ 11. pinocytosis	k. "little nut"
___ 12. autophagosome	l. movement apart
___ 13. oolemma	m. cell having few branches
___ 14. neurotropism	n. "form" of nucleus
___ 15. heterogamete	o. self-eating body
___ 16. kinetochore	p. true nucleus
	q. fat cell
	r. colored body
	s. slow movement

PART 2

Complete each of the scientific terms on the right with the appropriate missing part. Some terms are missing all parts!

1. one who studies cells	_____ ist
2. white blood cell	leuk_____
3. resembling a nucleolus	_____
4. natives before nuclei	pro_____
5. semifluid substance in muscle cell	_____ plasm
6. sheath around muscle cell	_____
7. formation of fat	_____ o _____ esis

8. nerve glue   neuro_____
9. dissolving body   _____ some
10. destruction of bacterial cells   _____
11. precursor egg cell   _____
12. sperm formation   _____ gen _____
13. affinity for myosin   _____ myosin
14. chromosome with equal arms   _____ chromosome
15. downward change   _____ bolism

ANSWERS

Part 1

1. i	9. m
2. n	10. c
3. r	11. e
4. a	12. o
5. p	13. d
6. b	14. f
7. k	15. g
8. q	16. l

Part 2

1. cytologist	9. lysosome
2. leukocyte	10. bacteriolysis
3. nucleoloid	11. ooblast
4. prokaryote	12. spermatogenesis
5. sarcoplasm	13. tropomyosin
6. sarcolemma	14. isochromosome
7. adipogenesis	15. catabolism
8. neuroglia	

UNIT 8

Chemistry

In this unit you will put together more than 60 terms of chemistry. You will use some of the parts and terms you already covered in previous units as well as the following:

an-, ana- *(up)*
cat- *(down)*
dis- *(opposite)*

acid/o *(acid)*
atm/o *(vapor, air)*
bas/i, /o *(base)*
calor/i *(heat)*
chem/o *(chemical)*
chromat/o *(color)*
electr/o *(electricity)*
equ/i *(equal)*
ion/o *(ion)*
kin/o, /e *(motion)*
neutr/o *(neutral)*
phil/e *(loving)*
phob/e *(fearing, hating)*

phor/e *(carrier)*
prot/o *(first)*
salin/o *(salt)*
stere/o *(three dimensions)*
therm/o *(heat)*
top/o *(place)*

-ate *(one acted upon)*
-ation *(action or process)*
-ify *(to make)*
-ism *(state, condition)*
-ity *(quality, state)*
-ization *(action or process)*
-ize *(to cause to be)*
-mer *(member of [specific] class)*
-on *(elementary particle)*
-sis *(process, action)*

chem

1. The science of chem/istry deals with the composition and properties of substances and with their transformations. The word root for chemistry is ⎯⎯⎯⎯⎯⎯⎯ .

kin/e
sis
kin/e/sis
process of action
or motion

2. Chemicals move about; the process is called kin/e/sis. Identify the two word parts:

_____/____ (combining form);

_____ (suffix).

Build it up again: _____/___/_____ .

What does it mean?

_____ .

kin
kə-NED-iks

3. The term for the branch of science that deals with the motion of substances uses the word root kin; it is called _____ /e/tics.

process of loosen-
ing up (breaking
up the whole into
its elements)

4. For chemists to determine the chemical structure of a substance, there must be an ana/lys/is (-sis) made. Study the word:

_____ (prefix);

_____ (word root);

_____ (suffix).

Build it up: _____/_____/_____ .

What does it mean?

_____ .

the breakup of
compounds into
their elementary
parts

5. A chemical ana/lys/is is_____

_____ .

chemical analysis

6. Analytical chemistry is that science that deals with

_____ _____ .

acid/i meter
acid/o id
instrument for
measuring acid
strength like acid

7. Many chemicals are acids or bases. To make acid into a combining form, it is necessary to add a vowel such as o, i, or u to the word. Circle the combining forms for acid in these two chemical terms:

acidimeter   acidoid

What does acidimeter mean?

_____ .

What about acidoid? _____ .

bears or contains
acid

8. A substance that is acid/i/fer/ous is one that
_____ .

bas/o
bas/i

9. The combining forms that mean chemical base are
_____/_____ and _____/____ .

10. Circle the combining forms for base in these
terms: basify   basicity   basoid

(bas/i)fy: to make
into a base
(bas/i)city: basic
state
(bas/o)id: baselike

What does basify mean?
_____ .

What about basicity?
_____ .

And basoid? _____ .

phil/e

11. The combining form that means loving is
_____/____ .

12. Build a term for something that is:
base-loving (attracted to bases)
_____ /o/_____/____ ;
acid-loving

bas/o/phil/e
acid/o/phil/e
hydr/o/phil/e

_____/____/_____/____ ;
water-loving
_____/____/_____/____ .

13. The suffix -ic makes these words adjectives. Write
them out as adjectives:

acid/o/phil/ic
bas/o/phil/ic
hydr/o/phil/ic

_____/___/_____/____ ;
_____/___/_____/____ ;
_____/___/_____/___ .

hydrophilic

14. Molecules that are attracted to water molecules
are _____ substances.

phil/ic

15. Atoms that are attracted to atomic nuclei and donate their electrons to them are nucle/o/ _____/____ atoms.

water-hating
(fear of water)

16. An acr/o/phob/ic person has a fear of heights. What does hydr/o/phob/ic mean?_____ .

hydr/o/phob/ic

17. Fats are not attracted to water molecules. They are _____/___/_____/_____ substances.

air- or vapor-
loving

18. Go to the list. Atm/o/phil/ic means _____ .

atmophilic

19. Certain chemical elements, like ozone, are found in or attracted to air. They are _____ molecules.

heat

20. Calor/i is one of the word-root combining forms that mean heat. A calor/i/meter is a device that measures absorbed or evolved _____ .

calorimeter
kal-ə-RIM-ə-dər

21. To measure the heat of combustion of coal or the change in temperature of melting ice, a researcher uses a _____ .

calor/i/gen

22. Build a word that means generating or forming heat _____ /i/_____ /ic.

calor/ie
KAL-ə-rē

23. Placing the ending -e onto the combining form for heat, you form the term _____ /ie.

calorie
kilo/calorie

24. The basic thermal unit for measuring heat in chemical reactions is the _____ .
Build a word that means a thousand calories: _____/_____ .

therm/o

25. Go back to the list. The other word-root combining form meaning heat is _____/____ .

heat

26. A therm/o/genic reaction is one that produces _____ .

therm/o

27. Mercuric oxide changes color when it is heated (from colorless at cool temperatures, to red at room temperature, to black when heated). Like many other substances, mercuric oxide exhibits _____/____ /chrom/ism.

instrument that
measures the heat
of water (or liquids
during reaction)

28. Analyze therm/o/hydr/o/meter:
_____/____ (combining form);
_____/____ (combining form);
_____ (word root).
Build it up: _____/_/____/_/_____
What does it mean?_____
_____ .

therm/al

29. Add the suffix -al and make the word root for heat into an adjective: _____/____ .

ex/o/therm/ic

30. Chemical reactions typically give off heat into or take in heat from the surroundings. Both kinds are therm/ic reactions. Build a word that means heat outside: _____/_/____/_____ .

exothermic

31. When methane burns, heat is evolved, or given off. The reaction is _____ .

end/o/therm/ic

32. Build a word that means heat within:

_____/_____/_____/_____ .

33. When calcium carbonate decomposes, heat
is absorbed from the surroundings. The reaction is

endothermic

_____ .

34. For both types of reactions, energy is always
conserved; that is, the amount of heat going out
equals the loss of internal energy in the reacting
substances. This internal property is given a name
that partly comes from the Greek word part thalp/y
(from *thalpein*, to heat), and partly from the prefix
meaning within, en-. Put the three word parts here:

_____ (prefix);

_____ (word root);

_____ (suffix).

en/thalp/y
EN-thal-pē
heat within

Build it: ____/_____/_____ .

What is its literal meaning?

_____ .

35. A general chemical rule is this:
Heat gained by the surroundings equals

____/_____/_____ lost by the

en/thalp/y
enthalpy

reaction; heat lost by the surroundings equals

_____ gained by the reaction.

36. Another property of chemical reactions
involves the quality of disorder and randomness
within the substances. Highly ordered substances,
like crystals, have a low level of randomness
or change; less ordered substances, like liquids
and gases, have a high level of randomness or

_____ . Using the word root meaning

change
en/trop

change, trop, build a word that means change

within: ____/_____ /y.

37. The general chemical rule is this: The _____/_____/_____ or "change within" of a substance increases as the temperature rises; _____ decreases as the temperature falls.

en/trop/y;
entropy

38. To review: The term that means heat within is _____/_____/_____ ; change within is _____/_____/_____ .

en/thalp/y
en/trop/y

39. An atom is made up of three major types of elementary particles. Using the appropriate root words and the suffix -on, build a word meaning:
first particle _____/_____ ;
electr/ical particle _____/_____ ;
neutr/al particle _____/_____ .

prot/on
electr/on
neutr/on

40. In short, the atom's three kinds of particles are the _____ , the _____ , and the _____ .

proton
electron
neutron

41. Build a word that means electr/on-loving (having an affinity for electrons):
_____/_____/_____ .

electr/o/phil/ic

42. An ion is a charged atom or group of atoms; that is, electrically charged atoms are _____ .

ions

43. Positively charged ions have fewer electrons than is necessary to remain neutral; negative ions have more. The names for these ions reflect the notion that electric current passes, like the sun, from east to west, from up to down (sunrise, sunset). The word ion comes from the Greek word *ienai*, to go. Look at the prefixes in this unit and then build a word that means: to go down _____/_____ ; to go up _____/_____ .

cat/ion
an/ion

cation

44. A cat/ion has fewer electrons than is necessary to remain neutral; that is, a positively charged ion is a _____ .

anion

45. An an/ion has more electrons than is necessary to remain neutral; in other words, a negatively charged ion is an _____ .

to cause to be an ion (to convert into ions)

46. Analyze ion/ize:
_____ (word root);
_____ (suffix).
What does it mean?
_____ .

to remove ions from (as water)

47. What does de/ion/ize mean?
_____ .

ion/ization

48. Build a word that means the process of ion/iz/ing:
_____/_____ .

the process of deionizing

49. What does de/ion/ization mean?
_____ .

generating, forming

50. An ion/o/gen is a compound capable of _____ ions.

phor/e
phor/e/sis
process of carrying

51. Write down the combining form meaning carrier:
_____/_____ . Now attach it to the suffix -sis.
Build a word from these parts:
_____/__/_____ .
What does the word that you formed mean?
_____ .

52. Build a word that means the process of carrying or transporting ions:
_____/__/_____/__/_____ .

ion/o/phor/e/sis
ī-än-ə-fə-RĒ-sis

53. The technique whereby ions are moved or transferred from one region to another is called _____ .

ionophoresis

54. The combining form is/o means equal. Attach it to -mer: _____/__/_____ .
What is this word's meaning?
_____ .

member of equal
class

55. An is/o/mer is a chemical compound with the same chemical formula as its sister compound *but* with different properties, owing to a different arrangement of atoms within this molecule. Ammonium cyanate, NH_4CNO, and urea, $CO(NH_2)_2$, are _____ .

isomers

56. A stere/o/scope is a device for making the images of two-dimensional pictures appear to have depth and solidity; in short, to appear in three dimensions. What is the combining form that means three dimensions?
_____/____ .

stere/o

57. The term for an isomer that has the same molecular formula as its sister but that has a different arrangement in three-dimensional space is
_____/__/_____/__/_____ .
(3-D)   (equal)   (part)

stere/o/is/o/mer

58. To is/o/mer/ize is to become changed into an isomeric form. Build a word that means the process of isomerizing:
_____/__/_____/_____ .

is/o/mer/ization

stereoisomer-
ization

59. Now build a word that means the process of stere/o/is/o/mer/iz/ing: _____ .

60. In many cases, by adding is/o to the name of a chemical, you can indicate an isomer of the chemical. Do that for these chemicals:

is/o   _____/___ /octane;
is/o   _____/___ /butane;
is/o   _____/___ /citric acid;
is/o   _____/___ /eugenol.

61. Is/o means equal or same. Using this combining form and the word part top/e (from the Greek *topos*, place), build a word that means same place:

is/o/top/e   _____/_/_____/_____ .

62. An is/o/top/e is one of two or more types of atoms that share the same chemical properties (they occupy the *same place* in the periodic table) but differ in physical properties. [14]Carbon is an _____ of

isotope   [12]Carbon.

63. The suffix -ic, makes a word an adjective. Using the suffix -ic, make is/o/top/e an adjective:

is/o/top/ic   _____/_/_____/_____ .

64. The number of neutrons minus the number of protons in an atomic nucleus is called the

isotopic   _____ number.

65. Analyze is/o/morph/ic:
_____/___ (combining form);
_____ (word root);
_____ (suffix).
Build it: _____/_/_____/_____ .

being of   What does it mean?
identical form   _____ .

66. A chemical having a similar or identical crystalline form as another (but with a different chemical composition) is an _____

isomorphic   substance.

67. Analyze is/o/di/morph/ic:
_____/____ (combining form);
_____ (prefix);
_____ (word root);
_____ (suffix).
Build it: _____/ /_____/_____/____ .
What does the word mean?
_____ .

Look it up.   (Now, look it up in your dictionary and see how close you came.)

68. Find the combining form that means electricity, again. Write it here. _____/____ .
Now combine it using the word part you
electr/o   built in frame 51, phor/e/sis. Put them together:
electr/o/phor/e/sis   _____/ /____/ /_____ .
process of carrying   What do you think it means?
through electricity   _____ .

69. In electr/o/phor/e/sis, suspended charged particles (ionized molecules) are moved by an electrical current to either side of a field depending on their electrical charge. When researchers wish to separate compounds out on the basis of their net electrical charge, they use
electrophoresis   paper or gel ____ _____ .

70. Write down the combining form that means color:
chromat/o   _____/____ .

71. Build a word that means record of color:
chromat/o/gram   _____/ /____/_____ .

chromat/o/gram

72. In chemistry, a chromat/o/gram is a record of the separation of substances by how well they dissolve in solvents (Latin: *solvere*, to dissolve). The separation is carried out on a paper strip, which is then stained. In other words, a series of compounds colored and separated on paper is called a _____/____/_____ .

chromat/o/graph

73. Now build a word that means recording device of color: _____/___/_____ .

chromat/o/graph/y
krō-mə-TAG-grə-fē

74. Now add the suffix -y to the term that you built: _____/___/_____/____ .

chromat/o/graph/y
chromatography

75. The process of separating gases, liquids and solids in solution on paper is called
paper _____/___/_____/____ ;
on a gel, it is called gel _____ .

equ/i

76. A triangle that is equ/i/angular has three equal angles. The combining form for equal in this term is _____/___ .

equ/i/caloric

77. Two compounds that have the same caloric values are _____/___/_____ .

equ/i/molar
equ/i/molal
equ/i/molecul

78. Use the combining form for equal to modify these terms:
having equal molar concentration
_____/___/_____ ;
having equal molal concentration
_____/___/_____ ;
containing an equal number of molecul/es
_____/___/_____ /ar.

equ/i/librium

79. The Latin word for balance is *libra*, from which comes the word part librium. Build a word that means equal balance: _____/_____/_____ .

equilibrium

80. A chemical reaction proceeds until there is a balance between the two sides of the reaction. It is then said to be in chemical _____ .

equ/i/valent
ə-KWIV-ə-lent

81. The Latin word *valens* means power or strength, from which comes the word part -valent. Build a word that means equal strength: _____/_____/_____ .

equivalent

82. The number of grams of an element that will combine with or replace one gram of hydrogen is called an _____ of that element.

to break
it down

83. To compose something is to build it up. What does de/compose mean?

_____ .

distill/ate
DIS-tə-lāt

84. The Latin word meaning to drip is *stillare*. To distill (de- + *stillare*) is to drip down or precipitate in drops, or to extract. The product of this process is called a _____/_____ .
            (drip down)   (one acted upon)

to separate

85. To associate is to join together. What does dissociate (dis- + associate) mean?

_____ .

dissociate

86. Under certain conditions, chemicals like sodium chloride (salt) _____ .

salin/o

87. Go to the list. The combining form meaning salt is _____/_____ .

salin/o/meter

88. Build a word for an instrument that measures salt: _____/___/_____ .

capable of forming salts

89. Salin/o/gen/ic means _____ .

salt

90. Salin/e water is water that contains _____ .

large molecule

91. A macr/o/molecule is a _____ .

many

92. Poly/atom/ic ions have _____ atoms.

beyond

93. A super/saturated solution is one that is _____ saturated.

centr/i/fugal

94. The Latin word meaning to flee is *fugere*, from which comes the word part -fugal. Build a word that means fleeing from the center: _____/__/_____ .

(center)   (flight)

centrifugal

95. When we twirl a bucket of water around on a long rope so that the contents are forced away from us, we are creating a _____ force.

centrifugal

96. A centr/i/fuge (SEN-trə-fyōōj) is a device that separates substances of different densities by using _____ force.

In this unit, you formed more than 60 new words. Thirty-one of these terms are listed below to allow you to practice your pronunciation. Pronounce each term several times before you take the Self-Test in this unit.

acidimeter	deionization	isomorphic
acidoid	deionize	isooctane
acidophilic	dissociate	kinesis
anion	electrophilic	nucleophilic
atmophilic	endothermic	neutron
basify	entropy	proton
basoid	equicaloric	salinometer
calorigenic	equilibrium	stereoisomer
centrifugal	ionophoresis	thermogenic
chromatogram	isomerization	thermohydrometer
decompose		

Unit 8 Self-Test

PART 1

From the list on the right, select the correct meaning for each of the following often-used chemical terms:

—— 1. acidimeter

—— 2. basicity

—— 3. hydrophilic

—— 4. atmophilic

—— 5. calorigenic

—— 6. thermochromism

—— 7. exothermic

—— 8. entropy

—— 9. enthalpy

—— 10. ionophoresis

—— 11. isotope

—— 12. cation

—— 13. salinogenic

—— 14. centrifugal

—— 15. deionize

a. generating heat

b. change within

c. process of carrying ions

d. heat within

e. salt-forming

f. basic state

g. negatively charged ion

h. water-loving

i. instrument that measures acid

j. acidlike

k. vapor-loving

l. water-hating

m. heat outside

n. remove ions

o. changes color upon heating

p. same place

q. flight from center

r. positively charged ion

s. add ions

PART 2

Complete each of the scientific terms on the right with the appropriate missing part. Some terms are missing all parts.

1. electron-loving   _____

2. negative ion   _____ ion

3. being of identical form _____ morph _____

4. color record of compounds _____ gram

5. equal strength   _____

6. containing acid   _____ i _____ ous

7. breaking up whole   ana_____

8. generating heat   _____ ic

9. first particle   _____

10. over saturated   _____ saturated

11. equal balance   _____

12. process of ionizing   ioniz_____

13. water-hating   hydro_____

14. science dealing with motion   _____ etics

15. measures heat of liquids   _____ hydr_____

ANSWERS

Part 1

1. i   9. d
2. f   10. c
3. h   11. p
4. k   12. r
5. a   13. e
6. o   14. q
7. m   15. n
8. b

Part 2

1. electrophilic   9. proton
2. anion   10. supersaturated
3. isomorphic   11. equilibrium
4. chromatogram   12. ionization
5. equivalent   13. hydrophobic
6. acidiferous   14. kinetics
7. analysis   15. thermohydrometer
8. thermogenic

UNIT 9

Biochemistry

In this unit you will put together over 80 terms of biochemistry by using many of the terms and parts you already covered (be sure that you are thoroughly familiar with them before beginning this section), plus the following:

aden/o *(gland)*
fruct/i *(fruit, fructose)*
galact/o *(milk, galactose)*
gluc/o *(glucose)*
glyc/o *(sugar)*
ket/o *(ketone)*
lact/o *(milk, lactate, lactose)*
nucle/o *(nucleic acid, nucleus)*
ox/y *(containing oxygen)*
phosph/o *(phosphoric acid)*
rib/o *(of or related to ribose)*

sucr/o *(sugar)*
thym/o *(thymus gland)*

-ase *(enzyme)*
-ide *(second, two parts)*
-idine *(chemical structure related
   to another compound)*
-il *(substance related to)*
-ine *(chemical substance)*
-ose *(sugar, carbohydrate)*
-t- *(third, three parts)*

life

1. From the organization of the term, bio/chemistry involves the study of the chemistry of

_____ .

gluc/o

2. Most cells use the simple sugar gluc/ose to provide energy. The word-root combining form for glucose is

_____ / _____ .

-ose

3. The suffix that indicates that gluc/ose is a sugar is

_____ .

sugar

4. Sucr/ose is found in a bowl on the kitchen table, in processed cereals, and in candy bars. Sucr/ose is table _____ .

formation of
glucose
glo͞o-ko̅-(genesis)

5. Gluc/ose can be taken in by the cell or made by the cell. When it is made, the process is called gluc/o/gen/e/sis, which means _____ .

gluc/agon

6. A hormone secreted by the pancreas causes the liver to release glucose sugar into the bloodstream. Part of the term for this hormone comes from the Greek word *agon,* driving or leading. Build a word that means driving sugar: _____/_____ .
                                                                (sugar)   (driving)

gluc/o/ne/o/
gen/e/sis

7. Sometimes cells must make glucose from the parts of other chemicals. Build a word that means formation of new glucose:
_____/__/_____/__/____/____ .
(sugar)   (new)   (formation of)

sugar or glucose

8. From an analysis of the term, the hormone gluc/o/cortic/oid must involve the metabolism of _____ .

glyc/o
glī-kə

9. When they are not using it, animal cells store glucose in the form of "animal starch" or glyc/o/gen. The word-root combining form for glyc/o/gen is _____/____ .

one that generates
sugar

10. Glyc/o/gen means _____
_____ .

glyc/o/lys
glī-KĂL-ə-sis

11. To get glucose sugar back from glycogen, glycogen must be broken down. Form a term that means the loosening or dissolving of sugar:
_____ / / _____ /is (-sis).

the dissolving of
sugar (what is
glucose?)

12. The process of the breakdown of glucose in the cell to make energy uses the same term. Why? What does the term glyc/o/lysis mean?
_____ .

glucose (sugar)

13. The glyc/o/lyt/ic pathway in the cell involves the breakdown of _____ for energy.

glyc/o/protein

14. Many proteins, such as those that act as cell membrane receptors, have sugars attached to them. Form a term that means sugar protein:
_____ / / _____ .

sugar

15. Analysis of glyc/o/phorin indicates that this molecule includes a _____ component.

fruct/i, /o

16. Another sugar is fruct/ose. Fruct/ose is often found in fruits; thus the name, fruit sugar. The word-root combining form for fructose is
_____ / ____ .

the breakdown
of fructose
frək-TĂL-ə-sis

17. Fruct/o/lysis means
_____ .

fructose

18. Circle the word-root combining form for fructose in phosph/o/fruct/o/kin/ase. The presence of the combining form in this term indicates that
_____ is involved.

-ase

19. Phosph/o/fruct/o/kin/ase is an enzyme. You can tell that this is the case by the suffix ＿＿＿＿＿＿ .

one that breaks down

20. Enzymes are proteins made by cells that act as catalysts in chemical reactions. What does cata/lys/t mean? ＿＿＿＿＿＿＿＿＿＿＿＿ .

enzyme

21. Most enzymes are highly specific in their action. They usually react with only one kind of molecule, called a substrate, to make a product. Enzymes are not consumed in the reaction. The reaction from substrate molecule to product molecule uses an ＿＿＿＿＿＿ .

layer below

22. Sub/strat/e means ＿＿＿＿＿＿＿＿＿＿ .

ase

23. While a few enzymes retain their old names (pepsin, trypsin), most enzymes are named by adding -ase to the word for the enzyme's substrate. Collagen is a protein found in most tissues. The enzyme that reacts with it is called collagen/＿＿＿ .

penicillinase

24. Build a term for the enzyme that reacts with penicillin: ＿＿＿＿＿＿＿＿＿＿ .

ase
ase
hī-DRÄJ-ə-nās
de/hydrogen/ase
dē-hī-DRÄJ-ə-nās

25. Enzymes are also named by adding the suffix -ase to the term for the type of reaction involved. Build a term for an enzyme involved in: the transfer of molecules from one location to another
transfer/＿＿＿ ;
an enzyme that attaches hydrogen to molecules
hydrogen/＿＿＿ ;
an enzyme that removes hydrogens
＿＿＿/＿＿＿＿＿/＿＿＿＿ .

(less   (hydrogen)   (suffix)
than)

26. Try these:

an enzyme that catalyzes isomerization

isomer/_____ ;

an enzyme that catalyzes a reduction (adds electrons)

reduct/_____ ;

an enzyme that fuses two molecules together
into one (Greek: *synthetos*, put together)

ase   synthet/_____ ;

ase

ase   an enzyme that catalyzes an oxidation (that removes

ase   electrons)

          oxid/_____ .

27. Try some more:

an enzyme that attaches oxygen molecules

_____/_____ ; an enzyme that activates
or deactivates other enzymes by adding a phosphate
group to them (that creates motion or action)

_____/_____ .

oxygen/ase   (motion,   (suffix)

kin/ase   action)

28. The terms for the enzymes that are named for
the jobs they do are coupled with the enzymes'
substrate names. The substrate name often precedes
the general name for the enzyme. What does
dihydrofolate reductase do? (What substrate molecule

adds electrons to   does it catalyze and how?)

dihydrofolate

          _____ .

synthesizes

glycogen   **29.** What does glycogen synthetase do?

(puts it together)

          _____ .

sugar   **30.** Galact/ose is a _____ .

         31. The combining form for the word that means

galact/o   milk sugar is g_____/_____ .

32. Build a term that means the enzyme that adds a phosphate group to galact/ose:

galact/o/kin/ase   ___/___/___/___ .

lact

33. The word root for lact/ose is _____ .

lact/ase

34. Build a term for the enzyme that catalyzes lact/ose: _____/_____ .

35. Build a word that means the enzyme that adds a phosphate group to gluc/ose: (Hint: Use the combining form for glucose).

gluc/o/kin/ase   ___/___/___/___ .

removes a hydrogen from glucose-6-phosphate

36. What is the job of glucose-6-phosphate de/hydrogen/ase? _____ .

removes a hydrogen from alcohol

37. What does alcohol de/hydrogen/ase do? _____ .

transfers adenylyl group from one location to another

38. How about adenylyl transfer/ase? _____ .

(phosph/o) fructo-kinase attaches a phosphate group to phospho-fructose

39. Circle the word-root combining form for phosphate in phosph/o/fruct/o/kin/ase. What does this molecule do? _____ .

ket/o

trans(ket/o)lase

40. Write the word-root combining form that means ket/one: _____/_____ . Locate and circle the combining form for ketone (or ketonelike) in trans/ ket/o/ lase.

transfers ketone groups

41. What does trans/ket/o/lase do?

_____ .

42. Proteins are made up of a number (sometimes hundreds) of small molecular units called amin/o acids. There are about 20 different kinds of amino acids. It is the sequence of different amino acids, like the letters of the alphabet, that creates the different kinds of proteins; that is, proteins are different because of their different _____

amino acid

sequences.

NH_2 or amino group

43. From their name, amin/o acids contain an _____ group.

peptides

44. A group of two or more amino acids strung together is called a peptide; that is, when proteins are broken up, the result is several and sometimes dozens of _____ .

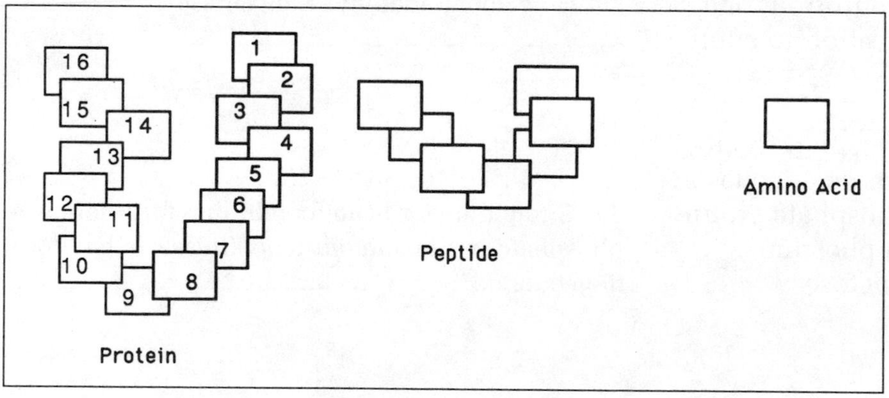

Protein

Peptide

Amino Acid

45. A di/peptide contains two amino acid molecules. How many amino acids is a pent/a/peptide made of?

five _____ .

46. How many amino acids does a dodec/a/peptide contain? _____ .

12

47. A poly/peptide contains _____ amino acids in its sequence chain.

many

48. Build a term that means an enzyme whose substrate is a peptid/e (and which breaks the peptide bonds between the different amino acids):

peptid/ase _____/_____ .

49. All cells contain nucle/ic acids. Most nucle/ic acids are used as molecules of information by the cell to build new cell parts. Analysis of the word root indicates that nucle/ic acids are, for the most part, found in the cell's _____ .

nucleus

50. Nucle/ic acids are made up of small units known as nucle/os/ide/s (nucle + ose + ide). Check the word part list. Analysis of the term indicates three things about the nucle/os/ide/s:

nucleic acid   1. _____ ;
contain a sugar   2. _____ ;
two-part molecule 3. _____ .

51. One part of a nucleoside molecule is a pent/ose molecule. Analyze the term. What two bits of information can you get from the term pent/ose?

five of something 1. _____ ;
it is a sugar   2. _____ .

52. The pentose molecule is a sugar that contains five carbon atoms. There are many kinds of pentose molecules. One pentose is called rib/ose. What is the combining form for rib/ose? _____/___ .

rib/o

53. Using the combining form for ribose, build a word that means a rib/ose nucle/o/side: ____/ /___/ /_____ .

rib/o/nucle/o/side

54. Ribose contains a full complement of oxygen atoms as well as its five carbon atoms. What is the combining form that means oxygen? _____/___ .

ox/y

55. Another pentose is identical to ribose except that it contains one less oxygen atom. Build a word that means a ribose with one less oxygen: _____/_____/_____/_____ .

de/ox/y/rib/ose   (less than)   (oxygen)   (ribose)

56. Now, build a word that means a de/ox/y/rib/ose nucle/o/side: _____/___ /_____/ /_____/ / .

de/ox/y/rib/o/
nucle/o/side

57. We now know that there are two kinds of nucleosides: one that contains the sugar ribose, the other that contains the sugar _____/___/_____/_____ .

de/ox/y/rib/ose

58. Linked to either of the two pentose sugars is one of any of five small molecules known as bases. Look in the word part list. Write the word root for gland: _____ .

aden

aden/ine

59. One base got its name from being extracted from glands. Using the suffix meaning chemical substance, build a word that means chemical substance from a gland: _____/_____ .

guan/ine
GWÄ-nēn

60. Another base is often found in guano, the Peruvian Indian name for dung. Remove the -o in guano and then build a term that means chemical substance from guano: _____/_____ .

cyt/os/ine
chemical sub-
stance from
sugar and cell

61. A third base is called cyt/os/ine (cyt + ose + ine). Analyze the term:

_____ (word root);
_____ (suffix fraction);
_____ (suffix).

Build it: _____/_____/_____ .
What does it mean?
_____ .

(Hint: Cytosine is often extracted from nucleic acids; they contain sugars and bases.)

substance related
to urea and acetic
acid YŌOR-ə-sil

62. A fourth base is ur/ac/il. Given this information:

ur/o	urea (urine)
ac	acetic acid
il	substance related to

what does this term mean?
_____ .

thym
thym/ine
THĪ-mēn

63. The final base is often extracted from the thymus gland of animals. What is the word root for thymus? _____ . Now build a word meaning chemical substance in the thymus: _____/_____ .

adenine
cytosine
guanine
uracil
thymine

64. Just to refresh your memory, name the five bases here:

_____ ; _____ ;
_____ ; _____ ;
_____ .

65. The five bases are grouped according to their general structure into two categories: cytosine, uracil, and thymine are called pyrimidines; adenine and guanine are called purines. Now rebuild your list of bases according to their categories:

Purines   Pyrimidines

_____   _____

_____   _____

                                            _____

66. The name of the molecule formed when a pyrimidine is attached to a pentose sugar is an alteration of the pyrimidine's name. To make such a name, remove everything except the first word root, and add the suffix -idine. Do that here:

idine   cyt/ _____ ;

idine   ur/ _____ ;

idine   thym/_____ .

67. The names of purines are also altered to name the purine-pentose molecules. For these names save the first word root, and add the word part -os/ine (ose + ine). What does -os/ine mean?

chemical
substance
of sugar   _____ .

68. Now, alter the names of the two purine bases to indicate that they are attached to pentoses:

os/ine   aden/_____/_____ ;

os/ine   guan/_____/_____ .

GWÄN-ə-sēn

69. To review, write the names for the bases, and next to them, the names of the same bases when attached to pentoses.

Purines

Bases   Altered Bases

_____   _____

_____   _____

Pyrimidines
Bases   Altered Bases

Check your
answers above.

70. We now know that a nucleoside consists of a base linked to a pentose. In a ribonucleoside, the pentose is ribose and the major nucleosides are adenosine, guanine, uridine, and cytidine. To form the major de/ox/y/ribonucleosides, you attach de/ox/y to each of the base names; also, you substitute thymidine for uridine. Do that here: deoxyadenosine;

deoxyguanine _____ ;
deoxycytosine _____ ;
deoxythymidine _____ .

71. When a phosphate group is attached to the pentose molecule of a nucleoside the nucle/o/side becomes a nucle/o/tide (t + ide); that is, when a phosphate group is attached to a nucleoside, you must indicate this with the suffix _____ .

-tide

72. Make rib/o/nucle/o/side into a nucle/o/tide:

rib/o/nucle/o/tide _____ .

73. Now make de/ox/y/rib/o/nucle/o/side into a nucleotide:

de/ox/y/rib/o/
nucle/o/tide _____ .

74. A nucle/o/tide is a single unit made up of a phosphate group, a sugar, and a base. Two chains of nucleotides make up the genetic information of the cell. Build a word that means many nucleotide units:

_____ .

poly/nucle/o/tide   (many)   (nucleotide)

75. The genetic information of most cells is in the form of two helical polynucleotide chains coiled around a common axis. The chains are made up of units of deoxyribose, bases, and phosphate groups. Build a name for the nucleic acid that contains de/ox/y/rib/ose using the word-root combining form for the sugar:

de/ox/y/rib/o/
(DNA)

_____/_____/_____/_____/_____ /nucle/ic
acid.

76. DNA is the molecule of heredity, made up of hundreds of thousands of deoxyribonucleotides. DNA is also a poly/mer. A poly/mer is a molecule made up of many repeating units. Build a word that means single unit: _____/___/_____ .

mon/o/mer

77. The DNA polymer is synthesized by an enzyme. Build a word that means enzyme that catalyzes DNA polymer building: DNA poly/mer/ _____ .

ase

78. Now, build a name for the nucleic acid that contains ribose: _____/___/_____/_____
acid.

rib/o/nucle/ic acid
(RNA)

79. Build a word that means enzyme that catalyzes synthesis of the RNA polymer:
RNA _____/_____/_____ .

RNA poly/mer/ase

80. The enzyme ribonuclease breaks up the RNA polymer into nucleotide units. Build a term for the enzyme that does the same thing to DNA:
_____ .

de/ox/y/rib/o/nucle/
ase

81. The genetic information in DNA must be read and then written into a molecular form that can leave the nucleus. When a secretary writes down the boss's words, the secretary is transcribing. The Latin word *scribere* means to write. What does trans/cribe mean?

to write across
(to make a copy)   _____ .

82. A transcript is a written copy. Build a word for the process of creating a written copy:

transcrip   _____ /tion.

83. In transcription, the information in DNA is copied onto a molecule of RNA. The next step is to translate this information into protein. *Latus* is the suppletive past participle of *ferre*, to bear or carry.

to carry across
(to change from
one place to
another)   Translate means _____ .

84. Now build a word that means the process of translating: _____ /tion.

transla

85. During translation, the information on an RNA polymer is used to create proteins. Amino acids are connected to each other on special bodies that contain rib/o/nucleic acids and proteins. Using the combining form for ribose, build a word for this structure, a term that means body of ribose:
_____ / _____ / _____ .

rib/o/some   (ribose)   (body)

86. Ribosomes are bodies made of RNA and protein. Using the combining form for nucleic acid, build a word that means nucleic acid and protein (as you did with glyc/o/protein): _____ / _____ / _____ .

nucle/o/protein

lip/o/protein

87. Some molecules are combinations of fats and proteins. Using the combining form for fat, lip/o, build a word that means fat and protein: _____/_____/_____ .

lip/ase

88. A lip/id is a kind of fat. Using the word root for this term, build a word that means an enzyme that is a catalyst of fat: _____/_____ .

phoph/o/lip/id

89. Now, build on the term lip/id. Form a word using the combining form for phosphate to build a term for a lipid that has a phosphate group attached: _____/_____/_____/_____ .

phosph/o/lip/ase

90. Form a term for the enzyme that catalyzes phosph/o/lip/ids: _____/_____/_____/_____ .

Here are 50 of the biochemistry terms you worked with in this unit. Be sure to pronounce each one carefully, and then complete the Unit Self-Test.

adenine	glycogen	polymerase
adenosine	glycolysis	polypeptide
catalyst	glycoprotein	reductase
cytidine	guanine	ribonuclease
cytosine	guanosine	ribonucleic
dehydrogenase	hydrogenase	ribonucleoside
deoxyribonuclease	isomerase	ribosome
deoxyribonucleic	kinase	substrate
deoxyribonucleotide	lactase	sucrose
deoxyribose	lipase	synthetase
fructose	nucleoside	thymidine
galactokinase	nucleotide	thymine
galactose	oxidase	transferase
glucagon	oxygenase	transketolase
glucogenesis	peptidase	uracil
gluconeogenesis	phosphofructokinase	uridine
glucose	phospholipid	

Unit 9 Self-Test

PART 1

From the list on the right, select the correct meaning for each of the following often-used biochemical terms:

____ 1. glycolysis
____ 2. fructose
____ 3. substrate
____ 4. transferase
____ 5. reductase
____ 6. oxidase
____ 7. kinase
____ 8. nucleoside
____ 9. deoxyribose
____ 10. guanine
____ 11. adenosine
____ 12. polymer
____ 13. transcription
____ 14. ribosome
____ 15. lipase

a. enzyme that moves molecules
b. nucleoprotein
c. copies DNA to RNA
d. many units
e. molecule of base and pentose
f. base of nucleoside
g. enzyme adds oxygen
h. enzyme catalyzes fat
i. enzyme removes electrons
j. fruit sugar
k. breakdown of glucose
l. enzyme adds electrons
m. under layer
n. base of nucleotide
o. pentose missing oxygen
p. copies RNA to protein
q. enzyme adds phosphate groups

PART 2

Complete each of the biochemical terms on the right with the missing part(s):

1. sugar protein   _____ protein
2. fusing enzyme   _____ ase
3. enzyme removes hydrogens   _____ ase
4. enzyme breaks up DNA polymer   _____ ase
5. enzyme synthesizes RNA polymer   RNA _____ ase
6. table sugar   _____
7. DNA   _____ acid
8. RNA   _____ acid
9. chemical substance from gland   _____
10. chemical substance from guano   _____
11. enzyme that adds oxygen   _____
12. peptide with 12 amino acids   _____ peptide

13. breakdown of fruit sugar   fruct_____
14. phosphate group on fat molecule   _____
15. five-carbon-atom sugar   _____

ANSWERS

Part 1

1. k	9. o
2. j	10. f
3. m	11. n
4. a	12. d
5. l	13. c
6. i	14. b
7. q	15. h
8. e	

Part 2

1. glycoprotein	9. adenine
2. synthetase	10. guanine
3. dehydrogenase	11. oxygenase
4. deoxyribonuclease	12. dodecapeptide
5. RNA polymerase	13. fructolysis
6. sucrose	14. phospholipid
7. deoxyribonucleic acid	15. pentose
8. ribonucleic acid	

UNIT 10

Physics

In this unit, you will put together more than 70 terms of physics. You will add new combining forms to the ones that you have already learned to build these scientific terms.

cath- *(down)*

audi/o *(sound, audible)*
bar/o *(pressure)*
chron/o *(time)*
dynam/o *(power)*
erg/o *(work)*
gravit/o *(gravity)*
gyr/o *(ring, spiral)*
kinet/o *(motion, movement)*
magnet/o *(magnetic)*
man/o *(gas, vapor)*

oscill/o *(swinging)*
phon/o *(sound)*
piez/o *(pressure)*
psychr/o *(cold)*
radi/o *(radiant energy)*
spectr/o *(spectra)*
stat/i *(motionless)*
synchr/o *(same time)*

-ode *(path)*
-tron *(device for manipulation of subatomic particles)*

electr/o/meter

1. Electr/o is used in words referring to electricity. An instrument that measures electricity (that measures electric-potential differences) is an

_____ / / _____ .

electr/o
electr/o/positive

2. Something that is charged with negative electricity is said to be _____ / ____ /negative; something charged with positive electricity is

_____ / / _____ .

electr/ic

3. Add the suffix -ic to the word root for electricity: _____/_____ . This makes the word an adjective.

electr/ic
electr/ic
electr/ic

4. The movement or current of positive and negative particles is called _____/___ current; the quantity of electricity held by a body is its _____/_____ charge; the region or field made by the closeness of electrons is called the _____/____ field.

electr/ode

5. Using the suffix -ode, meaning path, build a word meaning path of electricity: _____/___ .

cath/ode
an/ode

6. Using cath-, build a word for a negative (path) electr/ode: _____/_____ ; using an-, build a word for a positive (path) electr/ode: _____/_____ .

cathode

7. Recall that the suffix -on means elementary particle. An electr/on is a particle bearing a negative electric charge. Electrons are found at the (cathode/anode) _____ terminal.

cathode

8. A cathode ray is a stream of electrons emitted from a negatively charged electrode, or _____ .

oscill/o

9. Write the combining form for swinging: _____/___ .

oscill/ates
oscill/ation

10. When a pendulum swings back and forth, it _____ /ates. Add the suffix -ation to the word root for swing, to change it from a verb to a noun meaning the action of oscill/ating: _____/_____ .

oscill

11. A circuit designed to produce electric oscillations is an _____ /ation circuit.

oscill/o/graph

12. Using the combining form for swinging, build a word for the instrument that makes a record of periodic variations in electrical quantity: _____ / ___ / _____ .

   (swinging)   (record)

oscill/o/scope

13. A peri/scope is a device that is used to view around. Build a word for a device used to view electrical oscillations: _____ / ___ / _____ .

oscilloscope

14. A device that is used to view on a screen the waveforms made by electrons emanating from a cathode terminal is called a cathode ray _____ .

electr/o/lyte

15. The Greek word for loose or soluble is *lytos*, from which comes the word part -lyte. Using the combining form for electricity, build a word for a liquid conductor of electricity: _____ / ___ / _____ .

cath/o
cath/o/lyte
an/o/lyte

16. Form a combining form for the prefix cath-: _____ / ___ . Now build a word for the electr/o/lyte near the cath/ode: _____ / ___ / _____ What would be the term for the electr/o/lyte near the an/ode? _____ / ___ / _____ .

electr/o/motive

17. The word motive means moving to action. When you motiv/ate someone, you induce them into action. The force that moves electricity is an _____ / ___ / _____ force.

stat/ic

18. Something that is stat/ic, however, is motionless or at rest. Build a two-part term that means motionless electricity: _____/___ electricity.

stat/ic
stat/ic

19. The thrust developed by a jet engine at rest (with respect to the earth) is _____/___ thrust; the pressure exerted upon a surface at rest is _____/___ pressure.

electr/o/stat/ic

20. Using the term above and the combining form for electricity, build a word that means relating to stat/ic electricity: _____/___/___ ___/___ .

stat/ampere
stat/farad

21. In physics, stat often refers to electrostatic energy and is sometimes used to precede the names of electrical units. Ampere is an electrical unit for current. Make this term into an electrostatic unit: _____/_____ . Do the same with farad: _____/_____ .

kinet
kinet/ic

relating to the
motion of
electricity

22. Write the word root for movement _____ . Use it with the suffix -ic to build a word that means relating to the motion and forces of bodies: _____/___ .

23. Define electr/o/kinet/ic:
_____ .

instrument for
measuring
electrical motion

24. Analyze electr/o/kinet/o/graph:
_____/___ (combining form);
_____/___ (combining form);
_____ (word root).
Build it:
_____/ /___ / /_____ .
What does it mean?
_____ .
(Now, look the word up in your dictionary.)

focuses electrons
to see very
small things

25. A light microscope focuses light to view small objects. What does an electr/on microscope do? _____ .

26. Write the combining form meaning power: _____/___ ; use it to build a word meaning an

dynam/o/meter
dī-nə-MĂM-ə-dər

instrument for measuring mechanical (power) forces: _____/___/___ .

27. Using the combining form for power, build a word that means relating to the conversion of mechanical (power) energy into electrical energy: _____/___/___ /ic.

dynam/o/electr

    (power)   (electricity)

power

28. A dyne is a unit of _____ .

29. Analyze electr/o/dynam/o/meter:
_____/___ (combining form);
_____/___ (combining form);
_____ (word root).
Build it:

instrument for
measuring
electrical power

___/___/___/___ .
What does the word mean?
_____ .

30. A material that is magnet/ic has the ability to attract. What is the word-root combining form for this

magnet/o

word: _____/___ .

31. Build a word that means: a record of a magnetic phenomenon _____/___/___ ;

magnet/o/gram
magnet/o/graph
magnet/o/meter
mag-nə-TĂM-ə-dər

an instrument that records magnetic effects _____/___/___ ; an instrument that measures magnetic fields _____/___/___ .

32. Use the combining forms for magnet and for electricity to create two different terms with slightly different meanings: relating to electrical forces created by magnetic means _____/____/_____ /ic;

               (magnet)   (electricity)

relating to magnetic forces created by electrical

magnet/o/electr   means _____/____/_____/____ .

electr/o/magnet/ic   (electricity)   (magnet)

33. Analyze magnet/o/hydr/o/dynam/ics:

_____/____ (combining form);

_____/____ (combining form);

_____ (word root);

_____ (suffix).

Build it:

deals with motion   _____/____/_____/____/_____/____

of fluids in mag-   What do you think it means?

netic (and electric) _____ .

fields

(Hint: It is a branch of physics.)

34. Solar radi/ation is what keeps us warm and bathes the earth in light. The word root for this term

radi   is _____ .

35. The combining form for radiant energy, radi/o, is also the combining form that means radioactive. A radi/o/active particle emits radiant energy. Using the combining form for radioactivity,

build a term that means:

radi/o/carbon;   radioactive carbon _____/____/_____ ;

radi/o/isotope;   radioactive isotope _____/____/_____ ;

radi/o/sodium   radioactive sodium _____/____/_____ .

36. An opaque material is one that is impervious to visible light. A radi/opaque material is impervious to

radiation   _____ .

radi/o/lucent

37. A lucent material is as clear as glass and permits visible light to pass through. Build a term for a material that permits radiation to pass through: _____ / / _____ .

gravit/o

38. Gravit/y can be defined as the force between the earth and a body on its surface. The word-root combining form that means gravity is _____ / _____ .

gravit/o/meter
grav-ə-TÄM-ə-dər

39. Build a word for a device that measures the gravities of substances: _____ / / _____ .

erg/o

40. An erg is a unit of work. The combining form meaning work is _____ / _____ .

erg/o/graph;
erg/o/meter
er-GÄM-ə-dər

41. Using the combining form for work, form a term meaning:
a device that records work capacity _____ / / _____ ;
a device that measures work performed _____ / / _____ .

audi/o

42. An audi/o/phile is a person who loves (audible) sound. The combining form for audible sound (sound that we can hear) is _____ / _____ .

audi/o/gen/ic
audi/o/gram

43. Build a word that means:
generated or produced by audible sound waves _____ / /gen/ _____ ; record of auditory sound and vibrations _____ / / _____ .

spectr/o

44. When white light is passed through a prism, we see a spectr/um of the colors of the rainbow. The separation into colors is the result of light being resolved into its separate wavelengths and frequencies. What is the combining form that means spectr/um? _____/____ .

spectr/o/graph

45. A spectr/o/scope is an instrument used to view a spectrum (plural: spectr/a). Build a word that means an instrument that records spectra: _____/__/_____ .

a device that measures light

46. Define phot/o/meter: _____ .

a device that measures light spectra

47. Now define spectr/o/phot/o/meter: _____ .

spectr/o/pyr/o /meter
spectr/o/radi/o /meter

48. Build terms from word parts with the meanings: spectrum/fire (heat)/device that measures _____/_/___/_/_____ ; spectrum/radiation/ device that measures _____/_/_____/_/_____ . (Then look them up in your dictionary to understand their full meanings.)

measures sound over frequency ranges

49. Define audi/o/spectr/o/meter: _____ .

piez/o
bar/o

50. There are two combining forms that mean pressure. Look them up in the word part list in this unit and write them here: _____/__ ; _____/__ .

51. The combining form bar/o is used more often when referring to the atmosphere or to pressures of a grand measure, whereas piez/o is used when referring to more localized pressures. Choose the combining form you should use to build a word meaning a device that measures the pressure of the atmosphere: _____/_____ /meter.

bar/o
bə-RĂM-ə-dər

52. Now choose the combining form you should use to build a word meaning a device that measures the pressure of a liquid or solid: _____/_____ /meter.

piez/o
pē-ə-ZĂM-ə-dər

53. Keeping your two choices in mind, build a word meaning the science dealing with the effects of pressure on chemicals: _____/_____ /chemistry.

piez/o

54. A word used to define the mechanical pressures of large and heavy structures, such as bridges and dams would use the combining form _____/_____ . The term would be _____/_____ /dynamics.

bar/o
bar/o

55. The electricity due to pressure in a quartz crystal is called _____/_____ /electricity.

piez/o

56. The am/pere (AM-pir) is the unit of electrical current, named after the French physicist Andre Ampere. Knowing this, what do you think an am/meter is? _____ _____ .

a device that
measure electrical
current

57. When someone gyr/ates, they whirl around or move in a circle, ring, or spiral. The word root meaning spiral or ring is _____ .

gyr

gyr/o/magnet/ic

58. Using the combining form for spiral or spinning, build a word that means relating to (-ic) the magnetic (magnet) properties of a rotating electrical particle: _____/_/____/__ .

gyr/o/stat

59. Using the combining form for spinning and the word root meaning motionless, build a word that means a spinning device that stabilizes: _____/_/_____ .

gyr/o/scope
JĪ-rə-skōp

60. Build a word that literally means instrument for examining spins: _____/_/_____ .

gyroscope

61. A spinning wheel mounted to spin rapidly about any axis is called a _____ .

man/o

62. A man/o/meter is a device for measuring gaseous pressure. What is the combining form that means gas or vapor? _____/___ .

man/o/stat

63. Using the combining form for vapor and the word root for motionless, build a word that means a device for maintaining a constant pressure within an enclosure: _____/_/_____ .

chron/o

64. A chron/o/graph is a device for recording time. What is the combining form that means time? _____/___ .

chron/o/meter
krə-NÄM-ə-dər

65. Build a word that means a device that measures time: _____/_/_____ .

66. Build a word that means: instrument for
examining time _____/___/_____ ;
chron/o/scope   relating (-al) to both time and heat temperature
chron/o/therm/al   _____/___/_____/____ .

67. The combining form meaning sound or tone is
instrument for   phon/o. What is a literal definition of phon/o/graph?
recording sounds   _____ .

phon/o/meter   **68.** Build a word that means a device that measures
fə-NĂM-ə-dər   sound: _____/___/_____ .

69. Analyze phon/o/tel/e/meter:
_____/___ (combining form);
_____/___ (combining form);
_____ (word root).
Build it:
a device that   _____/___/____/___ .
measures distant   What does it mean?
sounds
_____ .

70. Analyze syn/chron/ous:
_____ (prefix);
_____ (word root);
_____ (suffix).
Build it:
_____/_____/____ .
happening at the   What does it mean?
same time
_____ .

71. The combining form that means same time is
_____/____ . Build a word that means
synchr/o   an instrument for examining events that occur
synchr/o/scope   together in time: _____/___/_____ .

72. Write down the suffix that means device for the manipulation of subatomic particles: _____ .
Now build a word that means a device that manipulates subatomic particles using electric and magnetic fields at the same time:

-tron

synchr/o/tron

_____ / / _____ .

(same time)   (device)

73. Psychr/o is the combining form that means

cold

_____ .

74. A device that uses two thermometers, one kept wet so that it is cooled by evaporation and another kept dry, so that one can measure the dryness of the air, is called a _____ / / _____ .

psychr/o/meter

sī-KRĂM-ə-dər

75. A device that makes a record of a psychrometer is called a _____ / / _____ .

psychr/o/graph

Here are 50 of the scientific terms you worked with in this unit. Pronounce each carefully, then go on to complete the Unit 10 Self-Test.

anode	electrometer	piezochemistry
anolyte	electrostatic	piezoelectricity
audiogenic	ergograph	piezometer
audiogram	ergometer	psychrograph
audiospectrometer	gravitometer	psychrometer
barodynamics	gyromagnetic	radiocarbon
barometer	gyroscope	radioisotope
cathode	gyrostat	radiolucent
catholyte	magnetoelectric	radiopaque
chronograph	magnetogram	spectrograph
chronometer	magnetometer	spectrophotometer
chronoscope	manometer	spectropyrometer
chronothermal	manostat	spectroradiometer
dynamoelectric	oscillograph	spectroscope
electrode	oscilloscope	synchroscope
electrolyte	phonometer	synchrotron
electromagnetic	phonotelemeter	

Unit 10 Self-Test

From the list on the right, select the correct meaning for each of the following often-used terms of physics:

____	1. electrometer	a. measures magnetic fields
____	2. electrode	b. permits radiation
____	3. oscilloscope	c. measures atmospheric pressure
____	4. cathode	d. of magnetic rotating particles
____	5. audiogram	e. maintains constant pressure
____	6. piezometer	f. positive electrode
____	7. radiolucent	g. varies with time
____	8. manostat	h. examines electrical waveforms
____	9. magnetometer	i. measures electricity
____	10. barometer	j. electricity "at rest"
____	11. gyromagnetic	k. radioactive particle
____	12. anode	l. electricity path
____	13. electrostatic	m. record of sound
____	14. dynamoelectric	n. negative electrode
____	15. radioisotope	o. stops radiation
		p. conversion of power to electric
		q. measures "small" pressures

Complete each of the scientific terms on the right with the missing parts. Some answers require you to give all parts.

1. electrolyte near negative terminal _____
2. charged with positive energy _____ positive
3. force that moves electricity   electr_____ ive
4. electrostatic unit of current _____
5. measures electrical motion _____ kin _____
6. record of magnetic phenomenon _____
7. radioactive carbon _____ carbon
8. measures gravities _____
9. records work capacity _____ o _____
10. instrument views spectrum _____

11. measures sound over frequencies
    we can hear   _____ o _____ o _____
12. measures gaseous pressure   m _____
13. measures time   _____
14. measures sound   _____
15. examines same-time events   _____

ANSWERS

Part 1

1. i		9. a
2. l		10. c
3. h		11. d
4. n		12. f
5. m		13. j
6. q		14. p
7. b		15. k
8. e		

Part 2

1. catholyte	9. ergometer
2. electropositive	10. spectroscope
3. electromotive	11. audiospectrometer
4. statampere	12. manometer
5. electrokinetograph	13. chronometer
6. magnetogram	14. phonometer
7. radiocarbon	15. synchroscope
8. gravitometer	

UNIT 11

Astronomy

In this unit, you will form more than 40 terms of astronomy. Some of them will be formed from the word parts you have picked up throughout the book. You will also use the following new prefixes, word-root combining forms, and suffixes:

pen- *(almost)*

anomal/o *(irregular)*
ap/o *(away from)*
aster/o *(star)*
astr/o *(star, the heavens)*
bar/y *(heavy)*
cosm/o *(universe, world)*
galact/o *(galaxy)*

heli/o *(sun)*
meteor/o *(meteor)*
selen/o *(moon)*
sider/o *(star)*
tel/e *(distant)*
tellur/o *(earth)*

-ite *(native)*
-oid *(resembling)*

astr/o

1. What is the combining form meaning stars or the heavens? _____/____ .

the stars and the heavens

2. Astr/o/nomy is the study of _____ .

3. The science dealing with the physical nature of heavenly bodies is _____/___/_____ .

astr/o/physics

(star)   (physics)

sailor
astr/o/naut

4. The Greek word for ship is *naus*. What do you think the word *nautes* means? Who lives and works on ships? _____ . Build a word using the word part for sailor, -naut, to mean one who travels in the heavens: _____/___/_____ .

astr/o/lith/o/logy

5. Lith/o is the combining form for stone. Build a word that means the science dealing with stones from the heavens (meteors): _____/___/_____/___/_____ .

has knowledge
of the laws of
the heavens

6. Examine the Greek word *astr/o/nomos*. *Nomos* is the Greek word for law. Give a literal definition of astr/o/nomer: one who _____ .

aster
aster/oid

7. Look in the list above. An aster/isk is a little star used in printing. What is the word root for star in this word? _____ . Use it to form a word that means resembling a star: _____/_____ .

planet/oid

8. Build a word that means resembling a planet: _____/_____ .

asteroids
planetoids

9. The thousands of bodies that are found in orbit around the Sun between Mars and Jupiter are called _____ or _____ .

tel/e

10. A tel/e/phone is a device for reproducing sound over a distance. Locate the combining form for distant: _____/_____ .

tel/e/scope

11. Build a word for the instrument that allows you to see distant objects: _____/___/_____ .

a device used to
see objects giving
off radiation

12. Define radi/o tel/e/scope:

_____ .

radi/o

13. A distant radio source that gives off strong
radiation is a _____/_____ star.

radi/o

14. The two-part term for the radio region of the sun's
electromagnetic spectrum is _____/_____ sun.

heli/o

15. Write down the combining form meaning sun:
_____/____ .

16. Analyze spectr/o/heli/o/graph:
_____/_____ (combining form);
_____/_____ (combining form);
_____ (word root).
Put it together:
_____/_/____/_/_____ .

device that records
the sun's spectrum

What does it mean?
_____ .

17. Analyze spectr/o/pyr/o/heli/o/meter:
_____ _____/_____ (combining form);
_____/____ (combining form);
_____/ ___ (combining form);
_____ (word root).

measures the
spectrum of
emitted radiation
from the sun

Build it:
_____/_/_____/_/_____/_/___ .

What do you think this device does?
_____ .

a device for taking
motion pictures
of the sun's
spectra over time

18. A kinemat/o/graph is a device for taking motion
pictures. What is a spectr/o/heli/o/kinemat/o/graph?
_____ .

moon

19. Selen/o/logy is the branch of astronomy that deals with the moon. Selen/o is the word-root combining form referring to the _____ .

selen/o/graph/y

20. Ge/o/graph/y is the science of the physical features of the earth. Build a term for the science of the physical features of the moon: _____/ / _____/ _____ .

selen/o/centr/ic

21. Centr/ic means relating to the center. Form a term that means relating to the center of the moon: / / _____/ _____ .

lunar

22. *Luna* is the Latin name for moon. From this comes the adjective lunar. What do you call the moon's day, or the period of rotation of the moon on its axis? _____ day.

stars and constellations

23. The Latin word for star or constellation is *sidus*. The word sider/eal comes from this and means of or relating to _____ .

sider

24. What is the word root meaning star or constellation? _____ .

sider
sī-DIR-ē-əl

25. The branch of astronomy that deals with the origins and relationships of the stars is _____ /eal astronomy.

sider/eal

26. The time in which the earth completes one revolution in its orbit around the sun with respect to the stars is called the _____/_____ year.

sider/o/graph

27. Build a term for the device that makes a record of the sidereal time: _____ / ___ / _____ .

sun

28. The Latin word for sun is *sol*, from which comes the adjective solar. A term containing this word refers to the _____ .

a telescope
designed to look
at the sun

29. Define solar telescope: _____ .

solar

30. An outburst or flare from the sun is called a _____ flare.

umbra

31. The word umbra means shadow. A body blocking out the sun's rays casts a shadow, or an _____ .

umbra

32. The dark central area of a sunspot is also called an _____ .

pen/umbra

33. Using the prefix pen-, build a word that means partial shadow: _____ / _____ .

pen/umbr

34. A partial eclipse of the moon is when the moon only passes through the earth's _____ / _____ /al shadow.

ap/o

35. Go back to the list. Write down the combining form meaning away from: _____ / ___ .

ap/o/gee

36. The word root for earth is ge. Build a word that means away from the earth (add an extra -e to ge): _____ / ___ / _____ .

apogee
AP-ə-jē

37. The point in the orbit of the moon (or other object) farthest from the center of the earth is called _____ .

peri/gee

38. The prefix peri- means near. Build a term meaning the point in the orbit of an object nearest to the center of the earth: _____/_____ .

heli/o/centr/ic

39. Using the Greek word-root combining form meaning sun, heli/o, build a word that means relating to the center of the sun: _____/___/_____/_____ .

peri/heli
ap/heli

40. Build a word that means:
the nearest point to the sun by a body orbiting it _____/_____ /on;
the farthest point from the sun by a body orbiting it _____/_____ /on.

ap/helion

41. When the earth is at its farthest point from the sun, it is at _____/_____ .

galact/o

42. A galaxy is one of billions of star systems. We live in the Milky Way galaxy. Write down the combining form for galaxy: _____/____ .

galact/o/centr/ic

43. Build a word that means relating to a galaxy as the center (of the universe). _____/___/_____/____ .

galact/ic

44. Add the suffix -ic to the word root of galaxy to make it an adjective: _____/____ .

galactic

45. The radiation coming from the Milky Way galaxy is called _____ noise.

46. An anomal/y is something that is out of place, a state of unevenness or irregularity. The combining form that means irregular is

anomal/o _____/_____ .

47. In astronomy, anomal/y refers to the position of a planet in its orbit. The anomal/y is the angle between the planet's perihelion position and its current position, using the sun as the center.

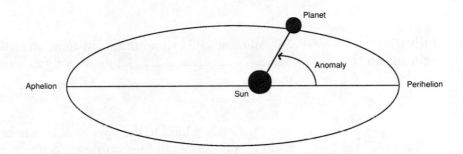

In other words, the angle between a planet and perihelion, as seen from the sun, is called the

anomaly _____ .

48. Add the suffix -istic to the word root of anomaly to make it an adjective: _____/_____ .

anomal/istic

49. The mean time of the moon's revolution from perigee to perigee again is called an

anomalistic _____ month.

50. The time of the earth's revolution from perihelion to perihelion again is about 365 days, or an

anomalistic year _____ _____ .

51. Write down the combining form meaning heavy:

bar/y _____/___ .

barycenter

52. The bar/y/center is the center of mass. When two suns revolve around each other, the _____ is somewhere between them.

bar/y/sphere
BAR-ə-sfir

53. Build a word for the heavy interior portion or sphere of the earth: _____/_____/_____ .

tellur

54. The Latin name for earth is *tellus*. Write the word root for earth: _____ .

tellur/ic
relating to the
earth

55. Add the suffix -ic to the word root for earth: _____/____ ; the word you wrote means _____ .

telluric
tə-LOO-rik

56. The bands added to the spectrum of a heavenly body by substances in the earth's atmosphere are called _____ lines.

cosm/o

57. Go back to the word list. The combining form that means world or universe is _____/____ .

universe

58. Cosm/o/logy is a branch of philosophy that deals with the character and origin of the _____ .

traveler of the
universe

59. Define cosm/o/naut: _____ .

cosm/o

60. The study of the chemical composition of the universe is called _____/____ /chemistry.

61. Using these parts:

cosm/o   (combining form)

tellur   (word root)

-ism   (suffix)

build a word that means something affecting both the heavens and the earth:

cosm/o/tellur/ism   _____/__/_____/___ .

62. Communication that is inter/planet/ary is

between planets   communication _____ .

63. A Latin word for star is *stella*, from which comes

stars   the adjective stellar, relating to _____ .

64. The passage of one star into the "shadow" of

stellar   another is called a _____ eclipse.

65. The energy of a star is called _____

stellar   energy.

66. Bodies that are inter/stellar move

between stars   _____ .

67. A meteor is a solid body from outer space. Some meteors survive the atmosphere to fall to the earth; the suffix -ite is used to distinguish them from those in space. Build a word for a meteor that has fallen

meteor/ite   to earth: _____/____ .

68. Given these two terms:

meteor/ist

meteor/itic/ist

which one means a person who studies

meteorist   meteors? _____ ;

meteoriticist   which one means a person who studies meteorites?

mēd-ē-ə-RID-ə-sist   _____ .

a very small   **69.** A micr/o/meteor/ite is
meteorite   _____ .

Here are 25 of the astronomical terms you worked with in Unit 11. Be sure to pronounce each one before taking the Unit 11 Self-Test.

anomalistic	heliocentric	sidereal
aphelion	interstellar	siderograph
apogee	meteoriticist	spectroheliograph
asteroid	penumbra	spectrohelio-
astrolithology	perigee	kinematograph
astrophysics	perihelion	spectropyrohelio-
barycenter	planetoid	meter
cosmotellurism	selenocentric	telescope
galactocentric	selenology	telluric

Unit 11 Self-Test

PART 1

From the list on the right, select the correct meaning for each of the following scientific terms:

___ 1. astrolithology	a. center of sun
___ 2. spectroheliograph	b. nearest to earth
___ 3. selenograph	c. of heaven and earth
___ 4. selenocentric	d. farthest from sun
___ 5. siderograph	e. spectrum of earth
___ 6. penumbra	f. nearest to sun
___ 7. apogee	g. of the earth
___ 8. heliocentric	h. one who studies meteors
___ 9. anomalistic	i. partial shadow
___ 10. barysphere	j. irregular
___ 11. telluric	k. record of sun's spectra
___ 12. cosmotellurian	l. record of moon
___ 13. perihelion	m. heavy core
___ 14. perigee	n. study of rocks from heavens
___ 15. meteorist	o. farthest from earth
	p. center of moon
	q. record sidereal time

PART 2

Write the scientific term for each of the following:

1. between stars _____
2. small meteorite _____
3. farthest orbit point from sun _____
4. study of the moon _____
5. star traveler _____
6. views distant objects _____
7. center of galaxy _____
8. farthest orbit point from earth _____
9. like a star _____
10. chemistry of universe _____

ANSWERS

Part 1

1. n	9. j
2. k	10. m
3. l	11. g
4. p	12. c
5. q	13. f
6. i	14. b
7. o	15. h
8. a	

Part 2

1. interstellar	6. telescope
2. micrometeorite	7. galactocentric
3. aphelion	8. apogee
4. selenology	9. asteroid
5. astronaut	10. cosmochemistry

Taxonomy

In this unit, you will form more than 60 new scientific terms. Some of these terms will be formed by using word parts you have learned from previous units. You will also use the following new root combining forms and suffixes.

amph/i *(both)*
angi/o *(vessel)*
arthr/o *(joint)*
aster/o *(star)*
brachi/o *(arm)*
brachy *(short)*
bry/o *(moss)*
cet/o *(whale)*
chil/o *(gill)*
chir/o *(hand)*
chondr/o *(cartilage)*
chord/o *(notochord)*
coel/e *(hollow cavity)*
cre/o *(flesh)*
crypt/o *(hidden, covered)*
cten/o *(comb)*
echin/o *(prickly)*
gastr/o *(belly)*
gnath/o *(jaw)*
gymn/o *(naked)*
lepid/o *(flake, scale)*
nemat/o *(thread)*

pelecy *(hatchet)*
phae/o *(brown)*
prot/o *(first in time)*
rhod/o *(red)*
schiz/o *(split)*
scyph/o *(cup)*
stomat/o *(mouth)*
tax/o *(arrangement)*
trache/o *(trachea)*

-acea *(animals characterized by)*
-aceae *(plants of the nature of)*
-ales *(plants belonging to)*
-ata *(oncs having)*
-fera *(group that bears)*
-ia *(taxonomic division)*
-ida *(animals that have the form of)*
-idae *(members of the family of)*
-idea *(animals that have the form of)*

-nomy *(sum of knowledge regarding a [specified] field)*
-phora *(organisms bearing a [specified] structure)*
-phyllum *(one having [such] leaves or leaflike parts)*
-phyta *(plants)*

-poda *(ones having [such] feet)*
-ptera *(organism[s] having [such] wings or winglike parts)*
-spermae *(ones having [such] a seed or germ)*
-vora *(ones that eat)*
-zoa *(animals)*

tax/o

1. The word-root combining form for arrangement is _____ / ____ .

arranges

2. A tax/i/derm/ist is one who _____ the skin of animals to create lifelike representations.

sum of knowledge

3. The word part -nomy means the sum of knowledge regarding a specified field. Agro/nomy is the _____ regarding fields or soil.

the sum of knowledge regarding arrangement

4. Tax/o/nomy, then, means _____ .

5. In tax/o/nomy, groups of plants and animals are arranged in a graded or ranked series that begins with the largest category and then divides into progressively smaller ones. The arrangement occurs this way:

Kingdom
    Phylum
        Class
            Order
                Family
                    Genus
                        Species

Now you write them out in order, largest to smallest:

_____

   _____

     _____

       _____

         _____

           _____

6. We belong to the following categories:
Kingdom: Animalia
   Phylum: Chordata
      Class: Mammalia
         Order: Primates
            Family: Hominidae
               Genus: *Homo*
                  Species: *sapiens*

Animalia

The largest category we belong to is
_____ (be specific).

sapiens

7. The smallest category is _____ .

8. The scientific name is always underlined or italicized, because it is a foreign name. For a particular animal or plant, the name is made up of the genus name (capitalized) followed by the species name. For example, the scientific name for the Canadian goose is *Branta canadensis*. The genus of

Branta

the Canadian goose is _____ _____ .

9. What would be the scientific name for humans?

Homo sapiens

_____ _____ .

10. Form the scientific name for the domestic dog from this information:
Family: Canidae
 Genus: *Canis*
  Species: *familiaris*

Canis familiaris   _____   _____ .

11. Try this one for the Norway rat:
Family: Muridae
 Genus: *Rattus*
  Species: *norvegicus*

Rattus norvegicus   _____   _____ .

poda

12. The word part that means ones having (so many) feet is _____ .

ones having
ten feet

13. Lobsters have ten limbs (feet); they belong to the order Deca/poda. What is the literal definition of this term?_____ .

joint

14. What does the combining form arthr/o mean? _____ .

Arthr/o/poda
är-THRAP-ə-də

15. Lobsters, insects, and scorpions all have jointed feet (appendages). Therefore, they all belong to the phylum _____/___/_____ .

Heter/o/poda

16. A group of snails has a foot that is different from those of other related groups. What is the term for this group that means different foot?
_____/___/_____ .

Gastr/o/poda

17. The combining form meaning belly is gastr/o. Form a term for the class of snails that have the foot by the belly: _____/___/_____ .

Oct/o/poda

18. What is the term for the order of animals that have eight feet? _____/___/_____ .

Cephal/o/poda

19. In squids and octopuses, the foot is wrapped around the head. Build a term for this class that means head-footed: _____/___/_____ .

Pelecy/poda

20. The clams, oysters, and mussels have feet shaped like hatchets. From the list at the beginning of the unit, build a word for this class that means hatchet foot: _____/_____ .

Arthr/o/gastr

21. Build a word for the division of animals that have jointed abdomens (bellies): _____/___/_____ /a.

Gastr/o/discoides

22. The Latin word for disk-shaped is *discoides*. Build a term for the genus of parasitic worms that invade the gut of man, the gut-discs: _____/___/_____ .

-idae

23. Generally, the suffix indicating that a term is in a taxonomic family is -idae, pronounced ə-dē. One family of snails is Helic/idae; a particular family of fish is Clupe/idae. The suffix that means "members of the family of" is _____ .

Equ/idae

24. Often the suffix -idae is substituted for the last syllable of the name of a genus to make a family term. Thus, *Aphis* becomes Aphid/idae. Build a term that means members of the family of *Equ/us* (horse): _____/_____ .

Lar/idae

25. Build a family term from the genus name for sea gull, *Lar/us*: _____/_____ .

Homar/idae

26. Now build one from the Genus name for the North American lobster, *Homar/us*: _____/_____ .

animals that have the form of scorpions

27. The suffix -ida means animals that have the form of. It is used mostly for the ranks of order and class. Scorpions belong to the order Scorpion/ida. What is the literal definition of this term? _____ .

group that bears perforations

28. A foramen (fə-RĀ-mən) is a perforation or small opening. What does the term foramen/i/fera mean? _____ .

Foramen/i/fer/ida

29. Build a term for the order of animals that have the form of foramenifera (remove the -a): _____/__/____/____ .

plants

30. The ending -phyta means _____ .

Chlor/o/phyta
Xanth/o/phyta
Chrys/o/phyta
Rhod/o/phyta

31. The term for the phylum of blue-green algae is Cyan/o/phyta. Build a phylum name for:
green algae _____/__/____ ;
yellow algae _____/__/____ ;
golden algae _____/__/____ ;
red algae _____/__/____ .

Phae/o/phyta

32. The combining form for brown is phae/o. Build a term for the phylum of brown plants (algae): _____/__/____ .

bry/o

33. What is the word-root combining form that means moss? _____/_____ .

Bry/o/phyta

34. Build a word that means moss plant, the phylum name for mosses, liverworts, and hornworts: _____/___/_____ .

myc/o

35. The word root for fungi is myc. What is the combining form? _____/_____ .

Myc/o/phyta

36. Build a word for the phylum name for fungi (fungi plants): _____/___/_____ .

trache/o

37. In humans, the trachea is a long tube that allows air to pass to and from the lungs. Ferns and seed plants have long, tubular cells that function in conduction of fluids; these cells are called tracheids. What is the combining form for trachea? _____/_____ .

Trache/o/phyta

38. Build a term for the phylum of vascular plants that possess these structures (trachea plants): _____/___/_____ .

-zoa

39. The suffix meaning animals is _____ .

moss

40. Bry/o/zoa is the taxonomic term for a group of animals that look like _____ .

Prot/o/zoa

41. Give the phylum name that means first (in time) animals: _____/___/_____ .

flower

42. Anth/o is the combining form meaning _____ . (Hint: botanical term.)

Anth/o/zoa

43. Coral and sea anemones are animals that look like flowers. Build a term for this class of animals: _____/___/_____ .

scyph/o
sīf-ə-

44. The true jellyfish appear as large cuplike animals. What is the combining form that means cup? _____/___ .

Scyph/o/zoa

45. Now, what is the term for the class in which these animals belong? _____/___/_____ .

Insect/i/vora

46. The ending that means ones that eat is -vora. The moles and shrews eat insects. They belong to the order _____ /i/_____ .

Carni/vora
kär-NIV-ə-rə

47. The Latin word for flesh is *carne*. To carni/fy is to turn into flesh. Build a term that means ones that eat flesh: _____/_____ .

notochord

48. Mammals, fish, reptiles, amphibians, and birds all have spinal columns or backbones which are preceded, during embryonic development, by a long rod called a notochord. The notochord gives all of these animals a common ancestry. They are all in the same phylum because they possess a _____ .

chord/o

49. What is the combining form for notochord? _____/___ .

Chord/ata

50. Use the suffix -ata, which means ones having, and the word root for notochord to build the term for the phylum name meaning ones having a notochord: _____/_____ .

Vertebr/ata

51. Birds, mammals, and reptiles are also vertebrates; they all have a column of back bones. Build a word meaning ones having vertebr/ae:
_____/_____ .

a backbone

52. A vertebrate has a backbone. What does an in/vertebrate not have? ____ _____ .

without

53. The In/vertebrata are animals (with/without) _____ backbones.

echin/o

54. The combining form that means prickly is _____/_____ .

Echin/o/derm/ata
ek-ə-näd-DER-
mə-də

55. Using the word root for skin, derm, build a term that means ones having prickly skin, a name for the phylum that includes starfish and sea urchins:
_____/__/_____/_____ .

coel/o

56. The combining form meaning hollow is _____/____ .

Coel/enter/ata
sē-len-tə-RÄ-də

57. The Greek term for intestine is _enteron_. A coel/enter/on is a hollow cavity. Build a term that means ones having a hollow cavity:
_____/_____/____ (jellyfish and coral).

-ptera
-p-tə-rə

58. What is the combining form that means organism(s) having (such) wing or winglike parts?
_____ .

Lepid/o/ptera

59. Use the combining form lepid/o (scale, flake) and build a word meaning organisms having scaly wings:
_____/__/_____ .

Lepid/o/ptera

60. Butterflies and moths have flakelike wings. What order do they belong to?

_____/___/_____ .

organisms having handwings

61. Refer to the word-root combining forms. What does the term Chir/o/ptera mean?

_____ .

Chir/o/ptera
kī-RĂP-tərə-ə

62. Bats have hands that are modified into wings. What order do they belong to?

_____/___/_____ .

organisms having two wings

63. What does the term Di/ptera mean?

_____ .

Di/ptera

64. Flies and mosquitoes have two wings. They belong to the order ____/_____ .

Cole/o/ptera

65. The combining form meaning sheath is cole/o. Beetles have thick front wings that cover their softer rear wings like sheaths. What order do they belong to? _____/___/_____ .

ales

66. The suffix -ales (ə-lēz) means plants belonging to. It is usually found at the taxonomic level of order. Many roses belong to the genus _Rosa_. Using the suffix -ales, build a word meaning plants belonging to the roses. Ros/_____ .

plants; order

67. In the term Rosales, the suffix indicates that it is a group of plants and that it is an order. In the term Chytridiales, what does the -ales signify?

_____ .

68. The suffix -aceae (ā-sē-ē) means plants of the nature of. It is used in the names of plant families. Thus, the term Acanth/aceae indicates

a plant family _____ .

Ros/aceae
rō-ZĀ-sē-ē

69. Make the genus *Ros/a* into a family term meaning plants of the nature of roses: _____/_____ .

-ales
-aceae

70. To review: The suffix _____ defines plants in an order. The suffix _____ defines plants in a family.

71. The suffix -acea (ā-shē-ə) means animals characterized by. It is usually used to indicate an order or class. Use the word root for whale to build a term for the order that includes whales and

Cet/acea

dolphins: _____/_____ .

72. The Latin word for shell is *crusta*. Build a word that means animals characterized by shells

Crust/acea

(use crust): _____/_____ .

73. Lobsters, crabs, shrimps, and barnacles all have hard shells. They belong to the class

Crustacea

_____ .

-idae
-acea

74. To review, the suffix _____ is used to indicate animal members of a family; the suffix _____ is often used to indicate animal members of an order or class.

-ida
-idea
-i-dē-ə

75. Look back in the suffix list. What are the two suffixes that mean animals that have the form of? _____ and _____ . Both are used to indicate animals of either a class or an order.

animals that have
the form of

76. We have used -ida previously.
Like -ida, the suffix -idea also means
_____ .

aster/o

77. The combining form for star is _____/_____ .

Aster/o/idea
astə-ROI-dēə

78. Using the suffix -idea, build a
word that means animals that have the form
of stars, the class that includes starfish:
_____/___/_____ .

Echin/o/idea

79. Build a term that means animals that have the
form of prickles, the class that includes sea urchins:
_____/___/_____ .

animals that have
the form of circle
leaves (an order of
tapeworm)

80. What is the literal definition of this term:
Cycl/o/phyll/idea?
_____ .

animals
a class or order

81. Generally, then, when you see -ida and -idea as
suffixes, they indicate (two concepts):
_____ ;
_____ .

does
(hidden gametes)

82. The suffix -ia (ē-ə), indicates that a class or order
of either animals or plants is involved. The term
Crypt/o/gam/ia (does/does not) _____
indicate either an order or a class of plants. (By the
way, can you figure out what the term means?)

Mammal/ia

83. Build a term for the class of mammals:
_____/_____ .

84. The ending -phyllum means one having (such) leaves or leaflike parts. The plants in the genus *Pod/o/phyllum* have leaves shaped like feet. What kind of leaves distinguish the plants in the group called Brachy/phyllum from the others?

short leaves

_____ .

85. Go back to the list. The word-root combining

amph/i

form for both is _____/_____ .

86. Amph/i often means that something does or is one thing and another. Amph/i/vorous means eating both animal and vegetable. Build a word that means ones having both (kinds of) feet: order

Amph/i/poda
(order of fleas)

_____/___/_____ .

ones having both
lives (leading
double lives)

87. What does the term Amph/i/bia (Greek: *bios*, life) mean? _____ .

88. The word part -spermae (SPER-mē) means ones having (such) a seed or germ. Use the combining form that means vessel, angi/o, to form a term for the class of plants that includes the flowering plants:

Angi/o/spermae

_____/___/_____ .

ones having a
vessel for their
seeds

89. What does the term Angiospermae mean?

_____ .

90. Using the combining form for naked, gymn/o, build a term for the class that includes the seed

Gymn/o/spermae

ferns: _____/___/_____ .

ones having
naked seeds

91. What does the term Gymn/o/spermae mean?

_____ .

brachi/o

92. Look up the combining form meaning arm. Write it here: _____/_____ .

Brachi/o/poda
brak-ē-ÄP-ə-də

93. Build a term that means ones having arm-feet: _____/___/_____ .

mouth

94. The word root for mouth is stomat. Stomat/o/logy is a branch of medical science concerned with the _____ .

Stomat/o/poda

95. Build a term that means ones having mouth-feet, for an order of Crustacea: _____/___/_____ .

Chil/o/stom/ata
kī-lə-STŌ-məd-ə

96. Using the combining form meaning gill, chil/o (kīl-ō), the word root for mouth, and the suffix -ata, build a term meaning ones having gill-mouths: _____/___/___/_____ .

gnath/o

97. The word root for jaw is gnath (nāth-ə). What is the combining form: _____/_____ .

Gnath/o/poda

98. Build a word meaning ones having jaw-feet, for the group of invertebrates whose jaws are modified limbs: _____/___/_____ .

Gnath/o/stom/ata

99. Build a taxonomic term using the combining form for jaw, the word root for mouth, and the suffix -ata for a group of animals having jaws: _____/___/___/_____ .

jaw

100. The word part -gnatha(e) sometimes is used to trail a term, but it still means _____ .

Chil/o/gnatha
kī-LAG-nə-thə

101. Using the combining form for gill and the trailing combining form for jaw, build a term meaning ones having a gill-jaw: _____ / / _____ .

chondr
kän-dr-

102. Go to the list. Write down the word root meaning cartilage: _____ .

Chondr/ichthyes
kän-DRIK-thē-ēz

103. The Greek word for fish is *ichthyes*. Using the Greek word for fish and the word root for cartilage, build a word for the class containing the sharks, skates, and rays—animals with cartilage instead of bone: _____ / _____ .

actually: winged
ones of cartilage
(skates, lampreys,
sharks)

104. What do you think the term Chondr/o/pterygii means? (Take a guess.) _____ _____ .

comblike

105. The combining form cten/o means comb. Organisms don't have combs per se; they do have comblike structures on or in their bodies. Organisms of the genus *Cten/o/plana* probably have _____ structures.

Cten/o/stom/ata
ten-ə-STŌ-məd-ə
(an order of
Bryozoa)

106. Build a taxonomic term using the combining form for comb, the word root for mouth, and the suffix -ata: _____ / / _____ / _____ .

Cten/o/phora
(a phylum)

107. Look in the list. Build a term that means organisms that bear a comb structure: _____ / / _____ .

nemat/o

108. The combining form for thread is nemat/o or nem/o. Nemat/oda is a class of minute threadlike roundworms. Many terms that describe threadlike animals or plants use the combining form
_____/_____ .

a family (-idae) of fish (ichthy) that are threadlike

109. Describe the organisms referred to by the term Nem/ichthy/idae:
_____ .

Nemat/o/gnathi

110. Using nemat/o and the trailing combining form for jaw, build a term for the order including catfish that literally means "thread jaw." (Replace the -a with -i.) _____/___/_____ .

Schiz/o/phora
skə-ZÄF-ər-ə

111. Schiz/o is the combining form meaning split. A schiz/oid is a person possessed of a split personality. Using this combining form, build a term that means organisms bearing a split or cleft:
_____/___/_____ .

Schiz/o/poda

112. Build a term that means ones having split feet:
_____/___/_____ .

Schizae/aceae

113. Build the term for the family name for the plant genus *Schizae*: _____/_____ .

In this unit you worked more than 60 new scientific terms. Fifty of those words are listed here for you to practice your pronunciation. Do that first, then take the Unit Self-Test.

Acanthaceae	Arthrogastra	Bryophyta
Amphipoda	Arthropoda	Bryozoa
Angiospermae	Asteroidea	Carnivora
Anthozoa	Brachipoda	Cephalopoda
Aphididae	Brachyphyllum	Cetacea

Chilostomata
Chiroptera
Chondrichthyes
Chordata
Coelenterata
Coleoptera
Crustacea
Cryptogamia
Ctenophora
Ctenostomata
Cyanophyta
Cyclophyllidea

Decapoda
Echinodermata
Echinoidea
Forameniferida
Gastropoda
Gnathopoda
Gnathostomata
Gymnospermae
Homaridae
Insectivora
Laridae
Lepidoptera

Mammalia
Mycophyta
Nematognathi
Nemichthyidae
Protozoa
Rhodophyta
Schizophora
Schizopoda
Tracheophyta
Vertebrata
Xanthophyta

Unit 12 Self-Test

PART 1

From the list on the right, select the correct meaning for each of the following taxonomic words.

_____ 1. order
_____ 2. Xanthophyta
_____ 3. Laridae
_____ 4. Anthozoa
_____ 5. Foramenifera
_____ 6. Echinodermata
_____ 7. Lepidoptera
_____ 8. Acanthaceae
_____ 9. Asteroidea
_____ 10. Arthropoda
_____ 11. Gymnospermae
_____ 12. Chilostomata
_____ 13. Chondrichthyes
_____ 14. Pelecypoda
_____ 15. Mycophyta

a. jointed foot
b. moss plants
c. yellow algae
d. ones having winged skin
e. immediately follows class
f. fungi plants
g. hatchet-foot
h. ones having gill-mouths
i. immediately precedes class
j. comb-animals
k. animals having form of stars
l. flower animals
m. head-foot
n. naked seeds
o. golden algae
p. a plant family
q. having scaly wings
r. member of animal family
s. ones having prickly skin
t. group that bears perforations
u. cartilage fish

PART 2

Complete each of the taxonomic terms on the right with the appropriate missing part. Some terms are missing all parts.

1. brown algae   _____ phyta
2. head-footed   _____ poda
3. animals having jointed bellies   Arthro_____
4. family name for *Homarus*
   (animal)   Homar_____
5. red algae   _____

6. moss plant _____ phyta
7. cup animals _____ zoa
8. one that eats insects Insecti _____
9. ones having a hollow cavity _____ enter _____
10. organisms having hand-wings _____
11. family name for *Rosa* (plant) Ros _____
12. animals that have the form of prickles _____ idea
13. hidden gametes _____ gamia
14. short leaves _____ phyllum
15. jaw-foot _____ poda
16. organisms that bear a comb structure _____

ANSWERS

Part 1

1. e
2. c
3. r
4. l
5. t
6. s
7. q
8. p
9. k
10. a
11. n
12. h
13. u
14. g
15. f

Part 2

1. Phaeophyta
2. Cephalopoda
3. Arthrogastra
4. Homaridae
5. Rhodophyta
6. Bryophyta
7. Scyphozoa
8. Insectivora
9. Coelenterata
10. Chiroptera
11. Rosaceae
12. Echinoidea
13. Cryptogamia
14. Brachyphyllum
15. Gnathopoda
16. Ctenophora

Review Sheets

Included in this section are the review sheets for each unit. The sheets will help you refresh your memory for the combining forms, suffixes, and prefixes that you worked on in *Quick Scientific Terminology*. You should feel free to rework these review sheets whenever you can so that you remember what you have learned in the text.

The answers are provided in the right-hand column; as always, you should cover the answer column with a card or a piece of paper while you work each sheet. The Index of Word Parts Learned in the back of the book will tell you where each word part was first introduced.

Unit 1: Review Sheet
Place, Direction, and Time

PART 1

WORD PART	MEANING	
de-	_____	from, down from
hypo-	_____	beneath
in-	_____	in, not
infra-	_____	under
inter-	_____	between
dextr/o	_____	right
di/a	_____	through
ant/e	_____	before, forward
anter/o	_____	before
caud/o	_____	tail
cephal/o	_____	head
circum-	_____	around
dors/o	_____	back
ect/o	_____	outer
end/o	_____	within
epi-	_____	over, upon
ex-	_____	from
ex/o	_____	outside
extra-	_____	beyond
hyper-	_____	above
ab-	_____	from
sub-	_____	below, under
trans-	_____	across, over
ventr/o	_____	belly
ultra-	_____	above
ad-	_____	toward
-al	_____	of, relating to
intra-	_____	within
later/o	_____	side
lev/o	_____	left
pre-	_____	before
pro-	_____	before, in front
post-	_____	after, behind
retr/o	_____	behind, backwards

sinistr/o	_____	left
super-	_____	beyond
supra-	_____	above
mes/o	_____	middle
medi-	_____	middle
mid-	_____	middle
para-	_____	around, near
per-	_____	through
peri-	_____	around
poster/o	_____	behind, after

PART 2

MEANING	WORD PART	
in, not	_____	in-
under	_____	infra-
between	_____	inter-
within	_____	end/o
side	_____	later/o
toward	_____	ad-
around	_____	circum-
behind, after	_____	poster/o
right	_____	dextr/o
through	_____	di/a, per-
back	_____	dors/o
before	_____	anter/o; pre-
tail	_____	caud/o
head	_____	cephal/o
from, down from	_____	de-
beneath	_____	hypo-
outer	_____	ect/o
over, upon	_____	epi-
from	_____	ab-
middle	_____	mes/o, medi-, mid-
around, near	_____	para-
before, in front	_____	pro-
after, behind	_____	post-
behind, backward	_____	retr/o
left	_____	lev/o, sinistr/o
beyond	_____	super-, extra-

above	_____	hyper-, supra-, ultra-
below, under	_____	sub-
across, over	_____	trans-
belly	_____	ventr/o
outside	_____	ex/o
of, relating to	_____	-al
before, forward	_____	ant/e

Unit 2: Review Sheet
Number, Size and Color

PART 1

WORD PART   MEANING

WORD PART		MEANING
bi-	_____	two, twice
undec/a	_____	eleven
xanth/o	_____	yellow
tri-	_____	three
atto-	_____	million trillionth
myri/o	_____	countless
mon/o	_____	one, single
uni-	_____	one, single
null/i	_____	no, none
demi-	_____	half
giga-	_____	billion
tera-	_____	trillion
peta-	_____	thousand trillion
micr/o	_____	minute, millionth
meg/a	_____	great, large
multi-	_____	many, much
hex/a	_____	six
chlor/o	_____	green
dec-	_____	ten
erythr/o	_____	red
semi-	_____	half
poly-	_____	many, several
centi-	_____	hundredth
pent/a	_____	five
melan/o	_____	black
chrys	_____	golden, yellow
pico-	_____	trillionth
hecto-	_____	hundred
exa-	_____	million trillion
chrom/o	_____	color
cyan/o	_____	blue
hemi-	_____	half
di-	_____	two, twice
femto-	_____	thousand trillionth

kilo-	_____	thousand
mega-	_____	million
tetr/a	_____	four
hept/a	_____	seven
leuk/o	_____	white
non/a	_____	nine
oct/a	_____	eight
milli-	_____	thousandth
nano-	_____	billionth
dodec/a	_____	twelve
deci-	_____	tenth

PART 2

MEANING	WORD PART	
trillion	_____	tera-
ten	_____	dec-
three	_____	tri-
blue	_____	cyan/o
color	_____	chrom/o
million	_____	meg/a
four	_____	tetr/a
seven	_____	hept/a
white	_____	leuk/o
red	_____	erythr/o
half	_____	demi-, semi-, hemi-
one, single	_____	mono, uni-
many, several	_____	poly-
hundredth	_____	centi-
billionth	_____	nan/o
twelve	_____	dodec/a
tenth	_____	deci-
black	_____	melan/o
golden, yellow	_____	chrys
thousand trillion	_____	peta-
six	_____	hex/a
green	_____	chlor/o
thousandth	_____	milli-
million trillion	_____	exa-
minute, millionth	_____	micr/o

many, much	_____	multi-
trillionth	_____	pico-
hundred	_____	hecto-
two	_____	bi-, di-
great, large	_____	meg/a
five	_____	pent/a
thousand	_____	kilo-
nine	_____	non/a
eight	_____	oct/a
eleven	_____	undec/a
yellow	_____	xanth/o
no, none	_____	null/i
million trillionth	_____	atto-
countless	_____	myri/o
billion	_____	giga-
thousand trillionth	_____	femto-

Unit 3: Self-Test
Basic Anatomy and Physiology

PART 1

WORD PART	MEANING	
cardi/o	_____	heart
cerebr/o	_____	brain
cost/o	_____	ribs
crani/o	_____	skull
-pnea	_____	breath, air
-or	_____	one that does
-ous, -ose	_____	full of
-scope	_____	instrument for viewing
-stasis	_____	checking
-tomy	_____	cutting
hepat/o	_____	liver
hist/o	_____	tissue
pneum/o	_____	air, breath
physi/o	_____	nature
ren/o	_____	kidney
splen/o	_____	spleen
tach/y	_____	rapid, speed
thorac/o	_____	chest
-tonia	_____	tonic spasm
troph/o	_____	nutrition
vas/o	_____	vessel
-ac	_____	pertaining to
derm/o	_____	skin
gangli/o	_____	ganglia
gen/o	_____	generate
gon/o	_____	semen, seed
hem/o	_____	blood
-ium	_____	small one, mass
-meter	_____	measuring instrument
-logist	_____	one who studies
-logy	_____	study of
-oid	_____	like, in the form of
-on	_____	unit
lip/o	_____	fat
a-	_____	from, away

a-, an-	_____	less, not
contra-	_____	against
dis-	_____	apart, not
em-, en-	_____	in, inside
eu-	_____	true
im-	_____	not
dys-	_____	difficult
acr/o	_____	extremity
blast/o	_____	formative cell
brad/y	_____	slow
morph/o	_____	form
my/o	_____	muscle
ne/o	_____	new, recent
nephr/o	_____	kidney
neur/o	_____	nerve
oste/o	_____	bone
-asis, -esis	_____	action, process
-ectomy	_____	cutting out
-gen	_____	generator
-gram	_____	record, write
-graph	_____	recording instrument
-ic	_____	of, relating to
-in	_____	belonging to

PART 2

MEANING		WORD PART
not	_____	im-
difficult	_____	dys-
extremity	_____	acr/o
formative cell	_____	blast/o
slow	_____	brad/y
heart	_____	cardi/o
brain	_____	cerebr/o
ribs	_____	cost/o
tissue	_____	hist/o
fat	_____	lip/o
from, away	_____	a-
less, not	_____	a-, an-
against	_____	contra-
nerve	_____	neur/o
skull	_____	crani/o

skin	_____	derm/o
ganglia	_____	gangli/o
generate	_____	gen/o
semen, seed	_____	gon/o
blood	_____	hem/o
liver	_____	hepat/o
bone	_____	oste/o
breath, air	_____	-pnea, pneum/o
small one, mass	_____	-ium
measuring instrument	_____	-meter
full of	_____	-ous, -ose
instrument for viewing	_____	-scope
checking	_____	-stasis
nature	_____	physi/o
kidney	_____	ren/o, nephr/o
spleen	_____	splen/o
rapid, speed	_____	tach/y
chest	_____	thorac/o
tonic spasm	_____	-tonia
one who studies	_____	-logist
one that does	_____	-or
nutrition	_____	troph/o
vessel	_____	vas/o
pertaining to	_____	-ac
apart, not	_____	dis-
in, inside	_____	em-, en-
true	_____	eu-
form	_____	morph/o
muscle	_____	my/o
new, recent	_____	ne/o
action, process	_____	-asis, -esis
cutting out	_____	-ectomy
generator	_____	-gen
record, write	_____	-gram
recording instrument	_____	-graph
of, relating to	_____	-ic
belonging to	_____	-in
study of	_____	-logy
like, in the form of	_____	-oid
unit	_____	-on
cutting	_____	-tomy

Unit 4: Review Sheet
Botany

PART 1

WORD PART	MEANING
cole/o	_____ sheath
rhiz/o	_____ root
coni/o	_____ dust (spores)
sapr/o	_____ rotten
-some	_____ body
herb	_____ grass
scler/o	_____ hard
xer/o	_____ dry
andr/o	_____ man
lign/i	_____ wood
anth/o	_____ flower
-phagous	_____ eating
phell/o	_____ cork
carp/o	_____ fruit
heter/o	_____ different
phot/o	_____ light
phyc/o	_____ seaweed, algae
-phyll	_____ leaf
clad/o	_____ shoot
phyt/o	_____ plant
coen/o	_____ shared in common
-plast	_____ formed
spor/o	_____ spore, seed
zyg/o	_____ yolk
bil/i	_____ bile
myc/o	_____ fungus
-idium	_____ little one
-cide	_____ killer
-vorous	_____ eating
-oid	_____ like
syn-, sym-	_____ together with
thall/o	_____ sprout
-trop	_____ turning
heli/o	_____ sun
homo	_____ same

PART 2

MEANING	WORD PART	
light	_____	phot/o
seaweed, algae	_____	phyc/o
leaf	_____	-phyll
shoot	_____	clad/o
rotten	_____	sapr/o
body	_____	-some
man	_____	andr/o
wood	_____	lign/i
flower	_____	anth/o
eating	_____	-phagous, -vorous
cork	_____	phell/o
fruit	_____	carp/o
little one	_____	-idium
killer	_____	-cide
plant	_____	phyt/o
shared in common	_____	coen/o
like	_____	-oid
grass	_____	herb
hard	_____	scler/o
spore, seed	_____	spor/o
yolk	_____	zyg/o
bile	_____	bil/i
fungus	_____	myc/o
dry	_____	xer/o
together with	_____	syn-, sym-
sprout	_____	thall/o
turning	_____	-trop
sun	_____	heli/o
same	_____	homo
formed	_____	-plast
sheath	_____	cole/o
root	_____	rhiz/o
dust (spores)	_____	coni/o
different	_____	heter/o

Unit 5: Review Sheet
Geology

PART 1

WORD PART	MEANING
cupr/o	_____ copper
all/o	_____ other
anis/o	_____ unequal
pale/o	_____ old
ped/o	_____ soil
aqu/a	_____ water
pel/o	_____ mud
auto-	_____ same
pyr/o	_____ fire
bio	_____ life
seism/o	_____ earthquake
calc	_____ lime
sider/o	_____ iron (ferrous)
carb/o	_____ carbon
silic/o	_____ silica
cry/o	_____ cold
strat/i	_____ layer
crystall/o	_____ crystal
uran/o	_____ uranium
ferr/o	_____ iron (ferric)
xen/o	_____ strange
fluvi/o	_____ river
-ite	_____ mineral, rock
ge/o	_____ earth
glaci/o	_____ glacier
hyal/o	_____ glass
hydr/o	_____ water
lith/o	_____ stone
metall/o	_____ metal
-fer	_____ one that bears
-aceous	_____ consisting of
-ic	_____ of, relating to

PART 2

MEANING	WORD PART	
of, relating to	_____	-ic
river	_____	fluvi/o
unequal	_____	anis/o
ferrous iron	_____	sider/o
cold	_____	cry/o
other	_____	all/o
water	_____	hydr/o, aqu/a
fire	_____	pyr/o
life	_____	bio
metal	_____	metall/o
layer	_____	strat/i
crystal	_____	crystall/o
carbon	_____	carb/o
silica	_____	silic/o
glass	_____	hyal/o
one that bears	_____	-fer
mud	_____	pel/o
same	_____	auto-
old	_____	pale/o
uranium	_____	uran/o
ferric iron	_____	ferr/o
copper	_____	cupr/o
glacier	_____	glaci/o
consisting of	_____	-aceous
stone	_____	lith/o
mineral, rock	_____	-ite
strange	_____	xen/o
soil	_____	ped/o
lime	_____	calc
earthquake	_____	seism/o
earth	_____	ge/o

Unit 6: Review Sheet
Organic Chemistry

PART 1

WORD PART	MEANING
para-	_____ two groups on 1, 4 positions on benzene
-oic	_____ carboxylic acid
but	_____ four carbon atoms
-ane	_____ saturated hydrocarbon
iod/o	_____ iodine
nitr/o	_____ nitrogen
meth	_____ one carbon atom
-yl	_____ branch group
brom/o	_____ bromine
-ol	_____ alcohol
eth	_____ two carbon atoms
amin/o	_____ amine group ($-NH_2$)
hydrox/y	_____ hydroxyl group ($-OH$)
-ine	_____ amine ($-NH_2$)
phenyl	_____ benzene with one hydrogen substitution
prop	_____ three carbon atoms
-ene	_____ double-bond hydrocarbon
-one	_____ ketone
-amine	_____ amine group ($-NH_2$)
fluor/o	_____ fluorine
chlor/o	_____ chlorine
ortho-	_____ two groups on adjacent carbon atoms on benzene
-yne	_____ triple-bond hydrocarbon
-al	_____ aldehyde
-ate	_____ carboxylic salt
meta-	_____ separated by one carbon atom on benzene

PART 2

MEANING	WORD PART
alcohol	_____ -ol
nitrogen	_____ nitr/o
one carbon atom	_____ meth
three carbon atoms	_____ prop
double-bond hydrocarbon	_____ -ene
fluorine	_____ fluor/o
triple-bond hydrocarbon	_____ -yne
aldehyde	_____ -al
chlorine	_____ chlor/o
two groups on adjacent carbon atoms on benzene	_____ ortho-
bromine	_____ brom/o
ketone	_____ -one
branch group	_____ -yl
two carbon atoms	_____ eth
hydroxyl group (–OH)	_____ hydrox/y
carboxylic salt	_____ -ate
separated by one carbon atom on benzene	_____ meta-
benzene with one hydrogen sub-stitution	_____ phenyl
two groups on 1, 4 positions of benzene	_____ para-
carboxylic acid	_____ -oic acid
four carbon atoms	_____ but
saturated hydro-carbon	_____ -ane
iodine	_____ iod/o
amine group (–NH₂)	_____ -amine, amin/o, -ine

Unit 7: Review Sheet
Cell Biology

PART 1

WORD PART	MEANING	
-plast	_____	formed
-plasm	_____	form
bacteri/o	_____	bacteria
centr/o	_____	center
gli/o	_____	glue
is/o	_____	equal
kary/o	_____	nucleus
ana-	_____	up, upward
anti-	_____	against
cyt/o	_____	cell
dendr/o	_____	like a tree
gamet/o	_____	gamete
o/o	_____	egg
phag/o	_____	eating
reticul/o	_____	net
-oides	_____	like
-osis	_____	formation
-ote	_____	native
vir/o	_____	virus
-some	_____	body
kinet/o	_____	movement
lemm/o	_____	shell, husk
lys/o	_____	loosening
mit/o	_____	thread
auto-	_____	self
cata-	_____	down, against
co-	_____	with, shared
adip/o	_____	fat
ax/o	_____	axon, axis
sarc/o	_____	flesh
spermat/o	_____	sperm
trop/o	_____	turning
nucle/o	_____	nucleus
olig/o	_____	few, small
-chore	_____	move apart

PART 2

MEANING	WORD PART	
like a tree	_____	dendr/o
gamete	_____	gamet/o
glue	_____	gli/o
equal	_____	is/o
fat	_____	adip/o
move apart	_____	-chore
flesh	_____	sarc/o
form	_____	-plasm
like	_____	-oides
formation	_____	-osis
axon, axis	_____	ax/o
bacteria	_____	bacteri/o
center	_____	centr/o
cell	_____	cyt/o
loosening	_____	lys/o
against	_____	anti-
self	_____	auto-
down, against	_____	cata-
with, shared	_____	co-
egg	_____	o/o
eating	_____	phag/o
net	_____	reticul/o
body	_____	-some
formed	_____	-plast
native	_____	-ote
thread	_____	mit/o
nucleus	_____	nucle/o
few, small	_____	olig/o
nucleus	_____	kary/o
movement	_____	kinet/o
shell, husk	_____	lemm/o
up, upward	_____	ana-
sperm	_____	spermat/o
turning	_____	trop/o
virus	_____	vir/o

Unit 8: Review Sheet
Chemistry

PART 1

WORD PART	MEANING
bas/i, /o	_____ base
calor/i	_____ heat
-ate	_____ one acted upon
-ation	_____ action or process
-ify	_____ to make
electr/o	_____ electricity
an-	_____ up
cat-	_____ down
dis-	_____ opposite
acid/o	_____ acid
atm/o	_____ vapor, air
equ/i	_____ equal
iont/o	_____ ion
kin/o, /e	_____ motion
neutr/o	_____ neutral
phor/e	_____ carrier
prot/o	_____ first
salin/o	_____ salt
-ism	_____ state, condition
-ity	_____ quality, state
-ization	_____ action or process
chem/o	_____ chemical
chromat/o	_____ color
stere/o	_____ three dimensions
therm/o	_____ heat
top/o	_____ place
-ize	_____ to cause to be
-mer	_____ member of (specific) class
-on	_____ elementary particle
phil/e	_____ loving
phob/e	_____ fearing, hating
-sis	_____ process, action

PART 2

MEANING	WORD PART	
ion	_____	iont/o
motion	_____	kin/o, /e
neutral	_____	neutr/o
carrier	_____	phor/e
chemical	_____	chem/o
heat	_____	therm/o
place	_____	top/o
one acted upon	_____	-ate
up	_____	an-
down	_____	cat-
color	_____	chromat/o
electricity	_____	electr/o
equal	_____	equ/i
action or process	_____	-ation
heat	_____	calor/i
elementary particle	_____	-on
loving	_____	phil/e
fearing, hating	_____	phob/e
to make	_____	-ify
state, condition	_____	-ism
three dimensions	_____	stere/o
quality, state	_____	-ity
action or process	_____	-ization
to cause to be	_____	-ize
member of (specific) class	_____	-mer
first	_____	prot/o
salt	_____	salin/o
opposite	_____	dis-
acid	_____	acid/o
vapor, air	_____	atm/o
base	_____	bas/i, /o
process, action	_____	-sis

Unit 9: Review Sheet
Biochemistry

PART 1

WORD PART	MEANING	
nucle/o	_____	nucleic acid, nucleus
ox/y	_____	containing oxygen
phosph/o	_____	phosphoric acid
-ine	_____	chemical substance
-ose	_____	sugar, carbohydrate
rib/o	_____	of or related to ribose
ket/o	_____	ketone
lact/o	_____	milk, lactate, lactose
-ase	_____	enzyme
gluc/o	_____	glucose
glyc/o	_____	sugar
-idine	_____	chemical structure related to another
-il	_____	substance related to
-ide	_____	second, two parts
sucr/o	_____	sugar
thym/o	_____	thymus gland
aden/o	_____	gland
fruct/i	_____	fruit, fructose
galact/o	_____	milk, galactose
-t-	_____	third, three parts

PART 2

MEANING	WORD PART	
milk, lactate, lactose	_____	lact/o
glucose	_____	gluc/o
fruit, fructose	_____	fruct/i
milk, galactose	_____	galact/o
phosphate	_____	phosph/o
of or related to ribose	_____	rib/o
sugar	_____	sucr/o, glyc/o

sugar, carbohydrate	_____	-ose
third, three parts	_____	-t-
ketone	_____	ket/o
second, two parts	_____	-ide
chemical structure related to another compound	_____	-idine
thymus gland	_____	thym/o
enzyme	_____	-ase
substance related to	_____	-il
nucleic acid, nucleus	_____	nucle/o
containing oxygen	_____	ox/y
gland	_____	aden/o
chemical substance	_____	-ine

Unit 10: Review Sheet
Physics

PART 1

WORD PART	MEANING
cath-	_____ down
audi/o	_____ sound, audible
bar/o	_____ pressure
chron/o	_____ time
dynam/o	_____ power
erg/o	_____ work
gravit/o	_____ gravity
gyr/o	_____ ring, spiral
kinet/o	_____ motion, movement
magnet/o	_____ magnetic
man/o	_____ gas, vapor
oscill/o	_____ swinging
phon/o	_____ sound
piez/o	_____ pressure
psychr/o	_____ cold
radi/o	_____ radiant energy
spectr/o	_____ spectra
stat/i	_____ motionless
synchr/o	_____ same time
-ode	_____ path
-tron	_____ device for manipulation of subatomic particles

PART 2

MEANING	WORD PART
gas, vapor	_____ man/o
swinging	_____ oscill/o
sound	_____ phon/o
spectra	_____ spectr/o
down	_____ cath-
sound, audible	_____ audi/o

same time	_____	synchr/o
path	_____	-ode
device for manipu- lation of sub- atomic particles	_____	-tron
pressure	_____	bar/o, piez/o
time	_____	chron/o
power	_____	dynam/o
work	_____	erg/o
gravity	_____	gravit/o
ring, spiral	_____	gyr/o
motion, movement	_____	kinet/o
magnetic	_____	magnet/o
motionless	_____	stat/i
cold	_____	psychr/o
radiant energy	_____	radi/o

Unit 11: Review Sheet
Astronomy

PART 1

WORD PART	MEANING	
meteor/o	_____	meteor
selen/o	_____	moon
astr/o	_____	star, the heavens
bar/y	_____	heavy
cosm/o	_____	universe, world
sider/o	_____	star
tellur/o	_____	earth
tel/e	_____	distant
pen-	_____	almost
anomal/o	_____	irregular
ap/o	_____	away from
aster/o	_____	star
galact/o	_____	galaxy
heli/o	_____	sun
-ite	_____	native
-oid	_____	resembling

PART 2

MEANING	WORD PART	
irregular	_____	anomal/o
away from	_____	ap/o
meteor	_____	meteor/o
moon	_____	selen/o
star	_____	aster/o, astr/o, sider/o
galaxy	_____	galact/o
sun	_____	heli/o
native	_____	-ite

resembling	_____	-oid
almost	_____	pen-
heavy	_____	bar/y
universe, world	_____	cosm/o
earth	_____	tellur/o
distant	_____	tel/e

Unit 12: Review Sheet
Taxonomy

PART 1

WORD PART	MEANING	
arthr/o	_____	joint
aster/o	_____	star
brachi/o	_____	arm
brachy	_____	short
bry/o	_____	moss
cet/o	_____	whale
-ida	_____	animals that have the form of
-idae	_____	members of the family of
-zoa	_____	animals
-acea	_____	animals characterized by
-aceae	_____	plants of the nature of
-ales	_____	plants belonging to
-vora	_____	ones that eat
crypt/o	_____	hidden, covered
cten/o	_____	comb
echin/o	_____	prickly
rhod/o	_____	red
gastr/o	_____	belly
gnath/o	_____	jaw
gymn/o	_____	naked
lepid/o	_____	flake, scale
nemat/o	_____	thread
pelecy	_____	hatchet
phae/o	_____	brown
prot/o	_____	first in time
-ata	_____	ones having
-fera	_____	group that bears
-ia	_____	taxonomic division
amph/i	_____	both
angi/o	_____	vessel
chil/o	_____	gill
schiz/o	_____	split
scyph/o	_____	cup

stomat/o	_____	mouth
tax/o	_____	arrangement
trache/o	_____	trachea
chir/o	_____	hand
chondr/o	_____	cartilage
chord/o	_____	notochord
coel/e	_____	hollow cavity
cre/o	_____	flesh
-idea	_____	animals that have the form of
-nomy	_____	sum of knowledge regarding a (specified) field
-phora	_____	organisms bearing a (specified) structure
-phyllum	_____	one having (such) leaves or leaflike parts
-phyta	_____	plants
-poda	_____	ones having (such) feet
-ptera	_____	organism(s) having (such) wings or winglike parts
-spermae	_____	ones having (such) a seed or germ

PART 2

MEANING	WORD PART	
red	_____	rhod/o
split	_____	schiz/o
cup	_____	scyph/o
group that bears	_____	-fera
ones that eat	_____	-vora
mouth	_____	stomat/o
arrangement	_____	tax/o
trachea	_____	trache/o
taxonomic division	_____	-ia
animals that have the form of	_____	-ida
one having (such) leaves or leaflike parts	_____	-phyllum

jaw	_____	gnath/o
naked	_____	gymn/o
flake, scale	_____	lepid/o
thread	_____	nemat/o
hatchet	_____	pelecy
plants	_____	-phyta
both	_____	amph/i
animals character- ized by	_____	-acea
plants of the nature of	_____	-aceae
plants belonging to	_____	-ales
ones having	_____	-ata
vessel	_____	angi/o
joint	_____	arthr/o
star	_____	aster/o
ones having (such) feet	_____	-poda

organism(s) having
   (such) wings or
   winglike parts
	_____	-ptera

ones having (such) a seed	_____	-spermae
animals	_____	-zoa
arm	_____	brachi/o
short	_____	brachy
moss	_____	bry/o
members of the family of	_____	-idae
prickly	_____	echin/o
belly	_____	gastr/o
brown	_____	phae/o
animals that have the form of	_____	-idea

sum of knowledge
   regarding a
   (specified) field
	_____	-nomy

organisms bearing a
   (specified)
   structure
	_____	-phora

whale	_____	cet/o
gill	_____	chil/o

hand	_____	chir/o
cartilage	_____	chondr/o
notochord	_____	chord/o
hollow cavity	_____	coel/e
flesh	_____	cre/o
hidden, covered	_____	crypt/o
comb	_____	cten/o
first in time	_____	prot/o

Final Self-Test I

INSTRUCTIONS

The next two tests are designed to let you know how well you have retained the information in this book. While some of the words on the tests will be new to you, most will not, and by using the word-building system, you should be able to decipher the meanings of the words presented here. Take your time and dissect each term carefully.

Each test includes 50 scientific terms gathered from all of the fields enclosed in this book. For each term, write out a definition for the term in your own words. The answers follow each test. Do not worry if your answer does not exactly match the book's answer. It is more important that you understand the definitions. Good luck!

TEST 1

1. tachycardia _____
2. apnea _____
3. hexadecagon _____
4. microcosm _____
5. electrochromatography _____
6. electromyograph _____
7. myriameter _____
8. carpogonium _____
9. creophagous _____
10. saprophagous _____
11. phycoerythrin _____
12. phycobilisome _____
13. mycophagist _____
14. microliter _____
15. xanthophyll _____
16. anthophore _____
17. neuroblast _____
18. myocardiograph _____

19. pentanomen _____
20. dorsolateral _____
21. endophyte _____
22. phonocardiogram _____
23. hyperventilation _____
24. hypodermal _____
25. superlunar _____
26. sinistromanual _____
27. dodecagon _____
28. electroencephalogram _____
29. dermatologist _____
30. craniocerebral _____
31. histogenic _____
32. osteocranium _____
33. pyroclast _____
34. nucleoprotein _____
35. atmophilic _____
36. thermohydrometer _____
37. fluviolacustrine _____
38. oosporiferous _____
39. allochemical _____
40. endothermic _____
41. fructolysis _____
42. dehydrogenase _____
43. electrostatic _____
44. penumbra _____
45. apogee _____
46. neurolemma _____
47. radiopaque _____
48. spectropyrometer _____
49. astrolithology _____
50. pedogenesis _____

ANSWERS TO FINAL SELF-TEST I

1. rapid heart
2. not breathing
3. sixteen sides
4. small world
5. separates substances with electricity
6. records electricity of muscles
7. 10,000 meters
8. fruit producing
9. eating flesh
10. eating decayed matter
11. red pigment of algae

12. bilelike pigment body of algae
13. one that eats fungi
14. millionth of a liter
15. yellow leaf pigment
16. one that bears flowers
17. formative nerve cell
18. instrument that records heart muscle
19. five-part name
20. toward the back and sides
21. plant living within another
22. record of heart sounds
23. excessive breathing
24. beneath the skin
25. above the moon
26. left-handed
27. twelve sides
28. measures electricity in brain
29. one who studies the skin
30. relating to skull and brain
31. generating tissue
32. bony skull
33. rock formed from fragments from a volcano
34. molecule of nucleic acid and protein
35. attracted to air
36. measures heat of liquids
37. formed by lake and river
38. bearing egg spores
39. other chemical
40. heat within
41. breakdown of fructose
42. enzyme that removes hydrogens
43. relates to static electricity
44. partial shadow
45. away from the earth
46. sheath (membrane) of nerve cell
47. impervious to radiation
48. measures spectrum of heat radiation
49. science dealing with meteorites
50. formation of soil

Final Self-Test II

See instructions on p. 254.

1. oligodendrocyte _____
2. cytolysosome _____
3. lipogenesis _____
4. hepatectomy _____
5. chlorocyte _____
6. tridecagon _____
7. rhizomorphous _____
8. xerophyte _____
9. coenobium _____
10. hyalocrystalline _____
11. cupriferous _____
12. silicification _____
13. carbonaceous _____
14. karyology _____
15. sarcoplasm _____
16. reductase _____
17. glucokinase _____
18. entropy _____
19. ionophoresis _____
20. isodecane _____
21. equicaloric _____
22. gluconeogenesis _____
23. octanomen _____
24. speleothem _____
25. leukocyte _____
26. nucleoplasm _____
27. chromosome _____
28. deoxyribose _____
29. DNA polymerase _____
30. acidiferous _____
31. hydrophilic _____
32. barycenter _____
33. electrokinetic _____
34. dynamometer _____

35. anion _____
36. oscilloscope _____
37. magnetograph _____
38. photomicrograph _____
39. Decapoda _____
40. Bryophyta _____
41. Echinodermata _____
42. Lepidoptera _____
43. gymnosperm _____
44. Chilostomata _____
45. Chondrichthyes _____
46. Schizophora _____

Draw out the carbon skeletons of the following:

47. 6-ethyl-3,3,5,5-tetramethylnonane _____
48. 1-bromo-2-chloro-3-iodobutane _____
49. 1,4-hexadiyne _____
50. 4-heptenoic acid _____

ANSWERS TO FINAL SELF-TEST II

1. cell with few treelike branches
2. cell dissolving body
3. formation of fat
4. cutting out the liver
5. green cell
6. thirteen sides
7. having the form of a root
8. plant living in a dry place
9. life shared in common
10. glassy-crystalline texture
11. bearing copper
12. replacement by silica
13. consisting of carbon
14. study of cell's nucleus
15. substance of muscle cells
16. enzyme that adds electrons
17. enzyme that adds phosphate group to glucose
18. change within; randomness
19. process where ions are moved from one region to another
20. isomer of ten-carbon alkane
21. with equal caloric values
22. formation of new glucose
23. eight-part name
24. cave mineral deposit
25. white blood cell
26. material in cell nucleus
27. colored body; genetic material
28. ribose minus one oxygen atom
29. enzyme catalyzes DNA polymer
30. contains acid
31. water-loving
32. center of mass
33. motion of electricity
34. instrument for measuring mechanical forces
35. negative ion

36. device to examine waveforms
37. records magnetic effects
38. image of a small object
39. having ten limbs
40. moss plant
41. those having prickly skin
42. organisms having scaly wings
43. naked seed
44. ones having gill-mouths
45. cartilage fish
46. organisms bearing a split

47.
```
         C—C   C       C
         |     |       |
C—C—C—C—C—C—C—C—C
         |     |
         C     C
```

48.
```
Br        I
|         |
C—C—C—C
   |
   Cl
```

49. C≡C—C—C≡C—C

50.
```
                        O
                        ‖
C—C—C=C—C—C—C—OH
```

Index of Word Parts Learned

The following word parts are listed by page number.

Additional Word Parts

Following are some word parts that you can use with your word-building system. If you want to expand your vocabulary with these word parts,

1. find a word part;
2. understand its meaning and context;
3. look it up in your dictionary;
4. make a list of words that have this word part in their makeup; and
5. develop a sentence that includes the word part so that the word part's meaning is clear to you.

WORD PART	MEANING	EXAMPLE
ailur/o	cat	ailur/o/don
-am	ammonia, related to	lact/am
amorph/o	formless	amorph/ism
anis/o	unequal	anis/o/trop/ic
anopl/o	unarmed	Anopl/anth/us
-ator	one that does	total/iz/ator
-atory	connected with	labor/atory
atr/o	black and	atr/o/castaneous
az/o	bivalent nitrogen	az/o/litmin
bath/o	depth	bath/ic
bdell/o	leech	Bdell/o/ura
bothr/o	trough, pit	Bothr/ops
-cace	diseased condition of a (type of) body part	arthr/o/cace
cac/o	bad, diseased	cac/o/chylia
-cade	spectacle	aqu/a/cade
call/o	beautiful	Call/o/rynch/us
calyc/o	calyx	calyc/oid
cari/o	caries	cari/o/gen/ic
-caris	shrimp	Echin/o/caris
-chroia	coloration	dys/chroia

-cnemic	shinned	platy/cnemic
-coelous	concave	pro/coelous
coen/o	common	coen/o/blast
coll/o	glue	coll/o/gen
cren/o	mineral spring	cren/ic
-ctonus	killer of insects	Dendr/o/ctonus
cub/o	cube	cub/o/manc/y
dacry/o	tears	dacry/oma
dactyl/o	finger, digit	dactyl/o/logy
e/o	earliest	e/o/lith/ic
ergat/o	worker	ergat/oid
eth/o	ethyl	eth/o/chlor/ide
eti/o	cause	eti/o/log/ic
fil/i	thread(s)	fil/i/fer/ous
hadr/o	thick, heavy	hadr/ome
hapl/o	single	hapl/oid
helic/o	helix, spiral	helic/o/graph
helminth/o	parasitic worm	helminth/o/logy
hel/o	marsh, nail	hel/o/phyte
idi/o	personal, distinct	idi/o/genet/ic
lei/o	smooth	lei/o/phyll/ous
lept/o	small, weak	Lept/andra
lyc/o	wolf	Lyc/o/pod/ium
malac/o	soft	malac/oid
malari/o	malaria	malari/o/logy
mer/a	sea, thigh, part	mer/maid; mer/algia
mer/i	part	mer/i/clin/ous
myrmec/o	ant	myrmec/o/logy
-myxa	resembling slime	Prote/o/myxa
myz/o	sucker	myz/o/dendr/on
nas/o	nose	nas/o/logy
necr/o	dead	necr/o/bi/o/sis
neph/o	cloud	neph/o/gram
ophi/o	snake	ophi/o/phag/ous
-opsis	having a part that resembles a	Chil/o/psis
-opsy	examination	bi/opsy
organ/o	organic	organ/o/tin
-orial	belonging to	gress/orial
osm/o	odor, smell	Osm/o/rhiza
-ost	bone	actin/ost
ostrac/o	shell	Ostrac/o/idea

ostre/o	oyster	ostre/oid
ot/o	ear	ot/o/logy
phac/o	lentil, eye	phac/o/meter
-phane	having a form	cym/o/phane
phaner/o	visible, open	phaner/o/cryst
pharmac/o	medicine	pharmac/o/logy
picr/o	bitter	Picr/o/dendr/o/n
pinn/i	feather, fin	pinn/al
pogon/o	beard	Pogon/ia
pteryg/o	wing, fin	pteryg/o/blast
ptyal/o	saliva	ptyal/agogue
-ptysis	spit, spittle	hem/o/ptysis
pycn/o	close, compact	pycn/ic
-pyga	having (such) a rump	Eury/pyga
rhipid/o	fan	Rhipid/o/istia
-rrhachis	spine	hemat/o/rrhachis
-rrhagia	abnormal flow	enter/o/rrhagia
-rrhea	flow, discharge	log/o/rrhea
-rrhine	having a nose	mon/o/rrhine
scat/o	excrement	scat/o/logy
schiz/o	split	schiz/o/gen/e/sis
scyt/o	skin	scyt/o/blast/ema
spec/i	species	spec/i/ation
spondyl/o	vertebra, whorl	spondyl/algia
sten/o	close, narrow	sten/o/mer/ic
streps/i	twisted	streps/i/tene
taut/o	same	taut/o/mer/ism
tot/i	whole	tot/i/palmate
tungst/o	tungsten	tungst/o/bor/ic
urin/o	urine	urin/o/logy
ur/o	tail, urine	ur/o/pod, ur/a/cil
-urus	tailed	Dasy/urus
vag/o	vagus nerve	vag/o/gram
-zoic	geological era, animal	Mes/o/zoic

NOTES

NOTES

NOTES

NOTES

NOTES